FRIED WATERS

EDUARDO DEL VALLE & MIRTA GÓMEZ

FRIED WATERS

NAZRAELI PRESS / JGS

INTRODUCTION

"the idea pure as salt" – St.-John Perse

Let us begin with the words of a poet because this book is a visual poem.

The photographers were raised in Cuba but live and work in Miami, Florida. They photograph on regular trips to Mexico, as if affirming a connection with their roots, but at a remove. Their photographs swim in a cultural Caribbean, which washes the shores of their native land, their adopted one and another which is connected in different ways with both.

These photographs were made in the Mexican province of Yucatán, a peninsular which juts towards Cuba and beyond that Florida. The pictures describe locations – at Xtambú and Dzilám de Bravo – in one of the most ancient, still-functioning, industrial sites in the world. Salt has been produced here since the Late Formative period of Mayan civilization, c. 300 – 50 B.C. Anthony P. Anderson, a historian of Mayan salt production and trade, describes the salt-beds of the northern coast of the Yucatán as "a veritable gift of nature to man, as they offer almost infinite amounts of this valuable resource in exchange for a limited amount of labor." The first Spaniard to describe the province, Bishop Landa writing in 1566, remarked that it produced "the best salt I have ever seen in my life, since it is very white when ground." Its whiteness guaranteed purity. Fray Antonio de Ciudad Real noted in 1588 that shiploads of salt were sent from Yucatán to Havana. The cover photograph of this book elegantly displays the working process: a salt-bed constructed of wood and impermeable materials – such as (in modern times) plastic sheeting – traps water in a naturally brackish lagoon. The "soft energy" of solar power evaporates the water during the dry season of January – June. The salt that remains is scooped from the bed with wooden hoes or shell scrapers into baskets such as the one seen in the salter's boat. The salt crystals are stacked in pyramids along the shore. Prior to transportation, the salt piles can be protected under thatch or glazed with torches, thus hardening them against even the heaviest rains.

Salt is both mineral and organic, medicinal and alimentary. It has been used in religious ritual, but also as money. Over 14,000 practical uses have been listed, including its role as an ingredient of the earliest photographic printing papers. Among its major applications are refining silver, tanning and, nowadays, the petrochemical industries. Salt solution is essential to the functioning of human blood, nerve impulses and heart action. However, in classical times Pliny described salt as necessary not purely for survival but for the *vita humanior*. Salt not only preserved such perishable items as fish but also brought out the savor in meats. It flavored cereals and contributed a vital supplement to vegetable-based diets. As the *Encyclopaedia Britannica* stated long ago, in the famous 11th edition published in 1913: "The habitual use of salt is intimately connected with the advance from nomadic to agricultural life and therefore with that step in civilization which had most influence on the cults of all ancient nations." Salt seems to be symbolic by its very nature. In Homer's *The Iliad* Achilles sprinkles "holy salt" on the roasted meat he offers to his gods and guests. "Ye are the salt of the earth," Jesus tells his disciples. St. Paul admonishes the Christians of Colossae near Galatia to "let your speech be always with grace, seasoned with salt."

The photographs here are seasoned with artistic knowledge. To photograph is, like other methods of depiction, simultaneously to observe and to imagine. The richer the imagination, the better the observation – and imagination is fed by the lore of any discipline. Eduardo del Valle and Mirta Gómez have an unusual working method: "We work in a coactive collaboration. We each use identical cameras (medium format 6x7cm cameras on tripods), shoot the same locations, mix-up the film and prints so that the authorship of single photographs is unverifiable. We print together (color negative film, enlarger, color processor, type-c prints) and edit from work prints. For the final cut – we each lobby in defense of our choices until we reach a consensus on behalf of the best photographs. This can take a few moments or a few years – sometimes, the arguments are more interesting than the photographic issues!" Years after working out their working method, they learned of the similar creative partnership of Bernd and Hilla Becher. Like the Bechers, they document structures. The presence of people is implied – and the implication is that the structures involve many generations not only our contemporaries. They are anonymous, generic and thus innumerably peopled.

Del Valle and Gómez initially worked, like the Bechers, in black and white. The first memorable photography book they saw and read was Garry Winogrand's *The Animals* (1969) with an essay by John Szarkowski. Their early experiments were mostly influenced by Garry Winogrand, Lee Friedlander, Eugène Atget, Diane Arbus and Walker Evans. Their first subjects were the streets and neighborhoods of Miami, all done with 35mm cameras. However, in 1984 they met the master of color photography William Eggleston. They visited him in Memphis and then invited him to give an informal presentation of his work at Florida International University. As they recently recalled, "We persuaded Bill to bring some prints for his presentation and he showed up with a trunk full of 5,000 photographs. As it turns out, he allowed us to study the prints for a few weeks." There could have been no better introduction to color photography. What did they gain from Eggleston? "He is a master at elevating the ordinary to extraordinary new heights with a *naturalness* and *ease* that conspire against all previous models in color photography. (With the exception of Helen Levitt's color work). His visual vocabulary is so broad and far reaching that it teaches us that 'all' things are worthy of attention or so it seems. This approach to art-making is powerfully inspiring and contagious and we enthusiastically applied what we learned from it in our own work, especially in Mexico."

Yucatán is the setting of four photographic projects, of which the first – *From the Ground Up*, on the evolution and metamorphosis of the vernacular domestic architecture of the region – was published in 2003. *Fried Waters*, the second volume in the series, will be followed by *Posthumous Portraits* – a survey of surviving bone fragments of Maya descendants, on their way to becoming dust. The final volume is titled *Down to Earth*, which traces the degeneration and regeneration of the earth's surface in the area of Chic-Xulub, where a meteor collided with the earth 65 million years ago. All the projects subtly interconnect.

Fried Waters begins with studies of ancient and abandoned salt pits whose poles now provide markers for small fishing boats. The second photograph includes, in the distance at the right, wild flamingos – descendants of the ones seen by Landa. Their pink plumage is caused by the crimson waters we shall see later in the book. The first section, *The Pit*, shows us abandoned workings. Some of the photographs appear to be elegantly simple descriptions of shape, construction and the unexpected colors the water can take on. Other photographs test our vision, asking us to tease out which parts of a pit are submerged and to differentiate between reflections (which are richly colored) and shadows (which are not). As we look more closely at some of these apparently straightforward photographs, we can detect the ways in which sunlight passes between the palisades around the pits and how it also strikes the bright edges of poles and bounces off at other angles. The locations in *The Pit* are filled with nets of light. Their micro-events are multiplied in the next section, *The Edge*. Here the photographers boldly use the technique of differential focusing to observe and symbolize the extraordinary events by which saline water becomes a thick brine, which hardens and crystallizes into salt. The photographs evoke this mysterious process with great skill. Finally, *The Lagoon*. Rarely have sheets of water been studied so carefully or rendered so precisely as in these photographs. The waters are landlocked and could be called stagnant, but they abound with organisms – including those that turn flamingo feathers pink. *Fried Waters* is a photographic poem about time and memory, place and labor, the symbolism of salt and the process by which it snaps from liquid into crystal. The poet Pablo Neruda described the same thing in his poem "Salt":

> I saw the salt
> in this shaker
> in the salt flat
> I know
> you will never believe me
> but it sings
> the salt sings.

I think we believe both the photographers and the poet.

Mark Haworth-Booth
Visiting Professor of Photography,
University of the Arts London
Honorary Research Fellow, Victoria and Albert Museum

Note: I am indebted to Eduardo del Valle and Mirta Gómez for a lively and illuminating email correspondence in 2004. I gratefully acknowledge my debts to recent writers on salt, including S.A.M. Adshead, Anthony P. Andrews, Pierre Laszlo and Mark Kurlansky.

FRIED WATERS

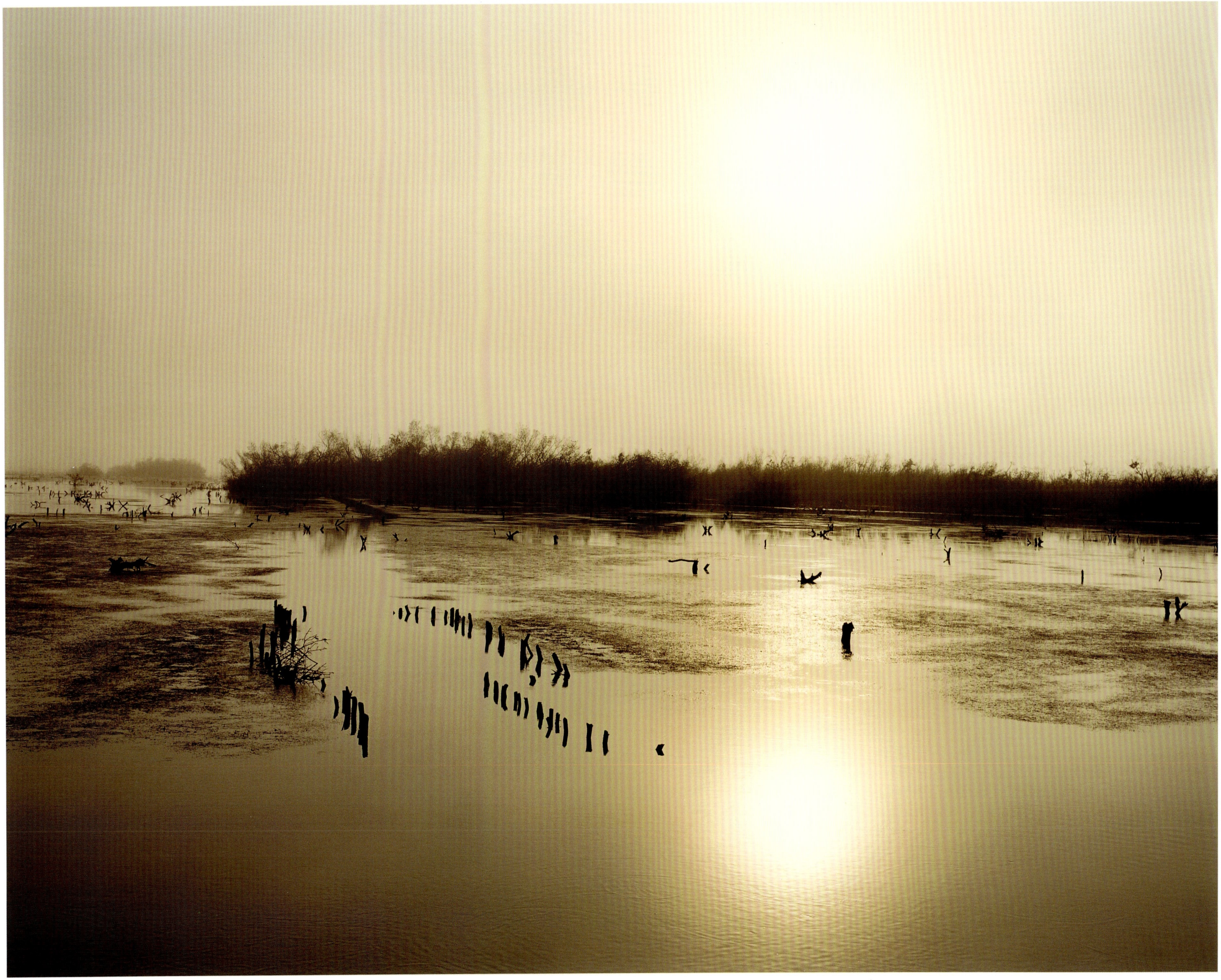

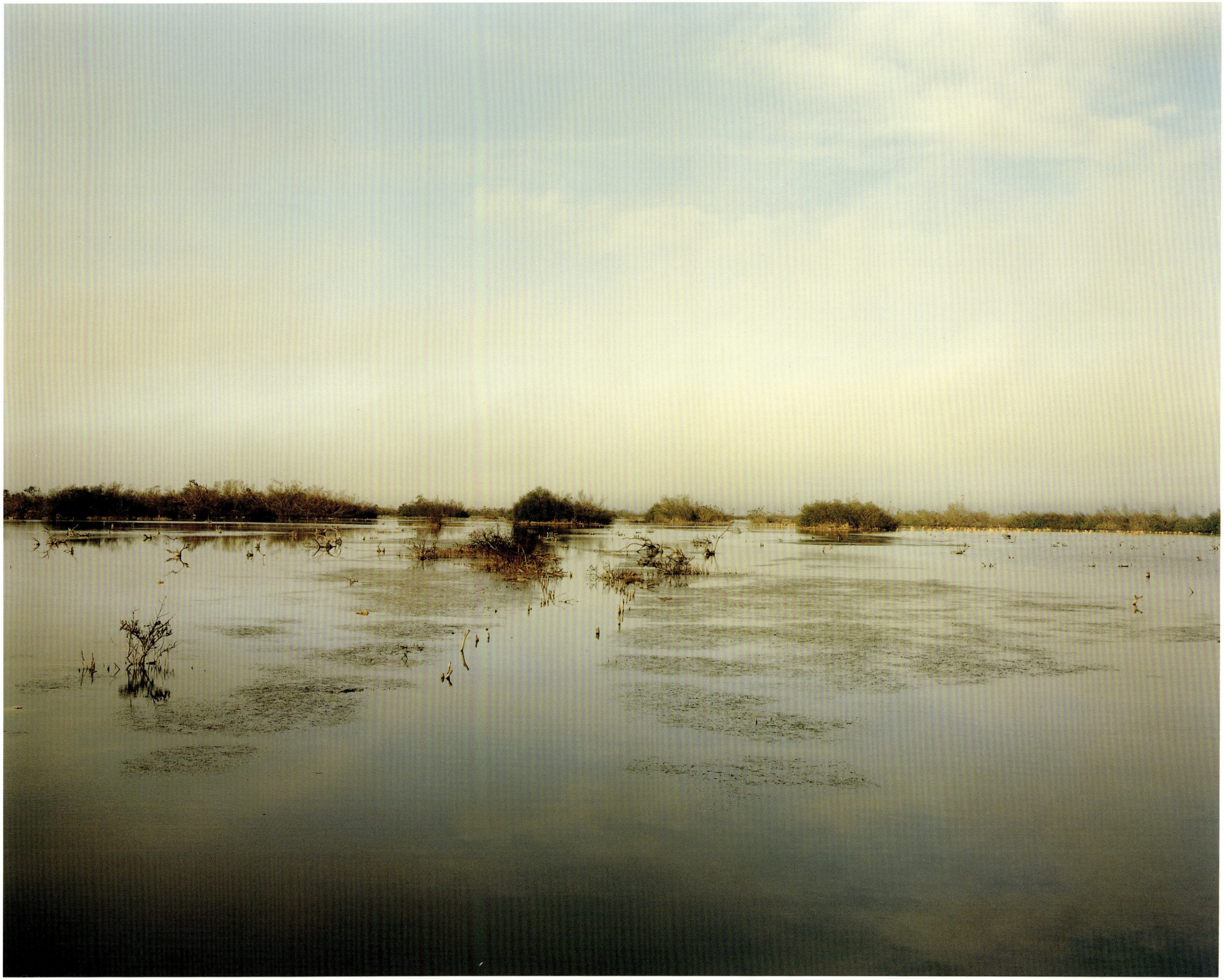

THE PIT

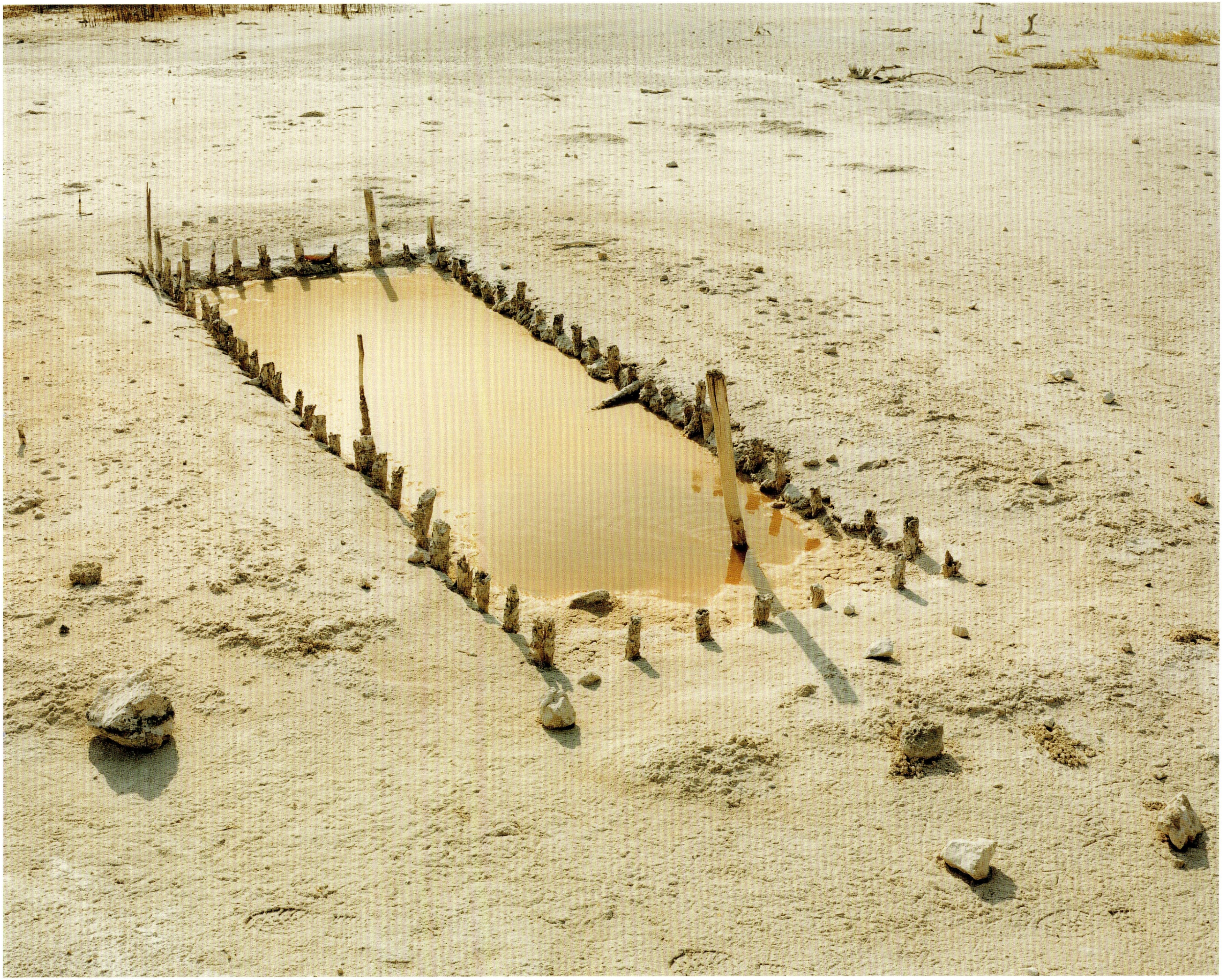

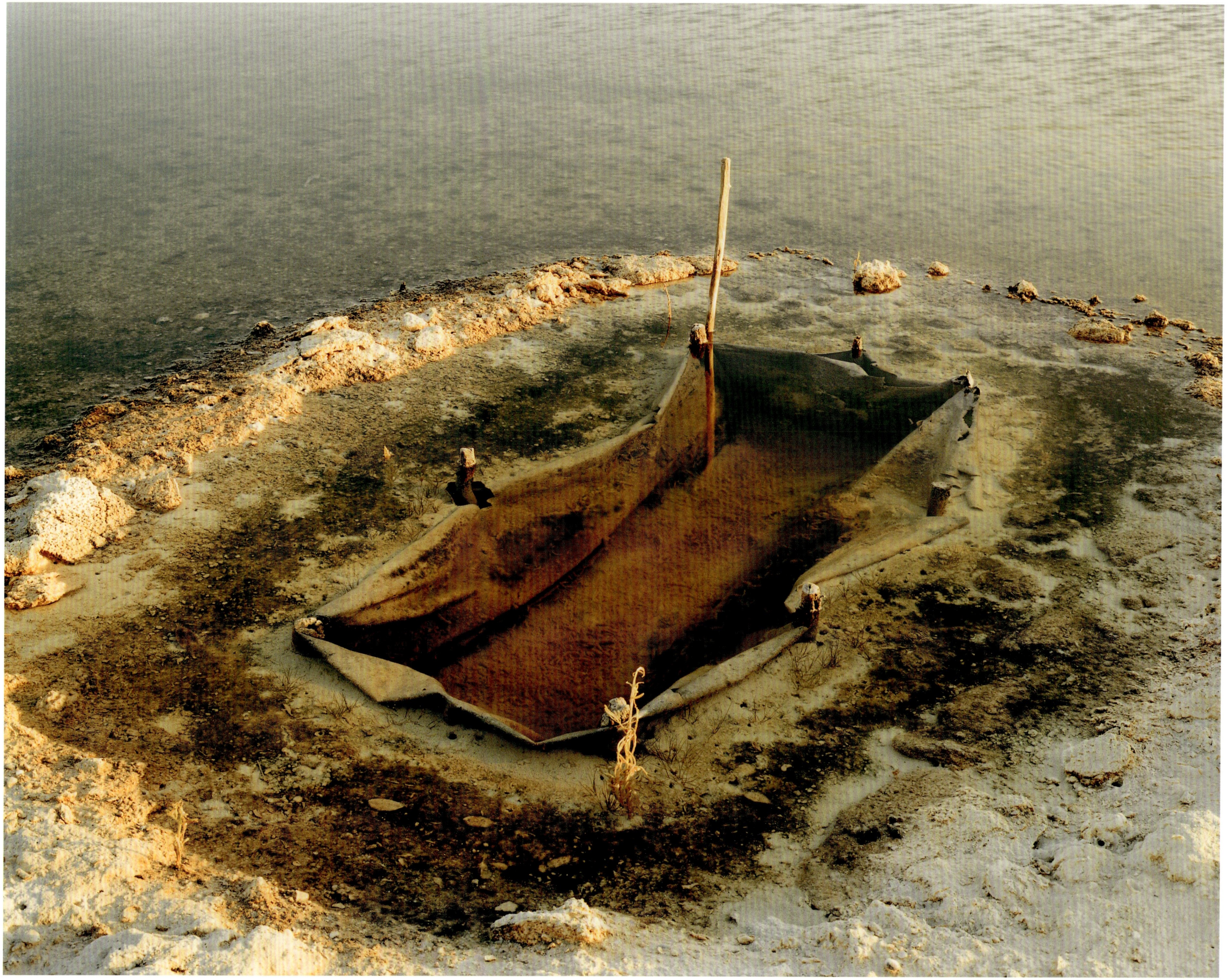

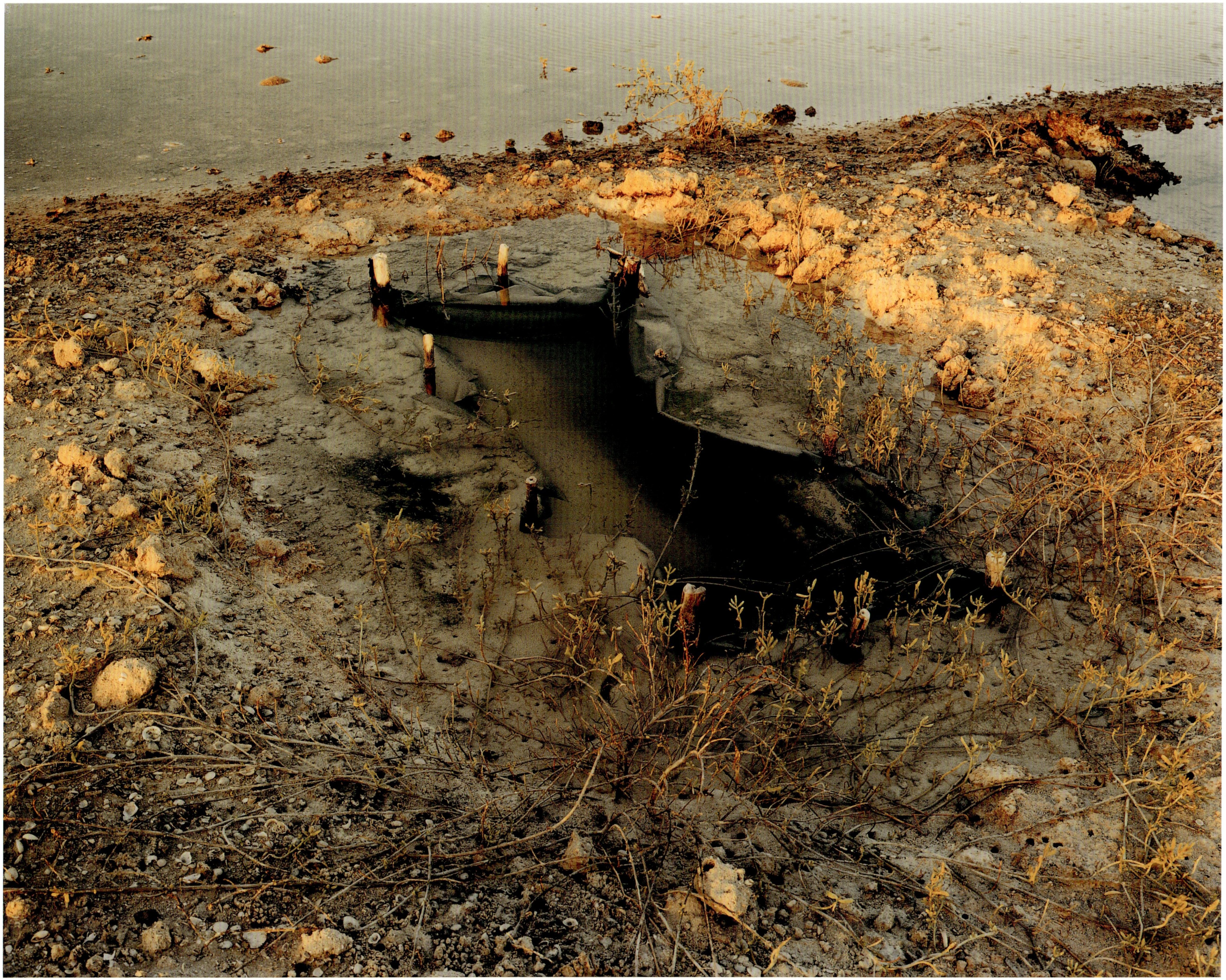

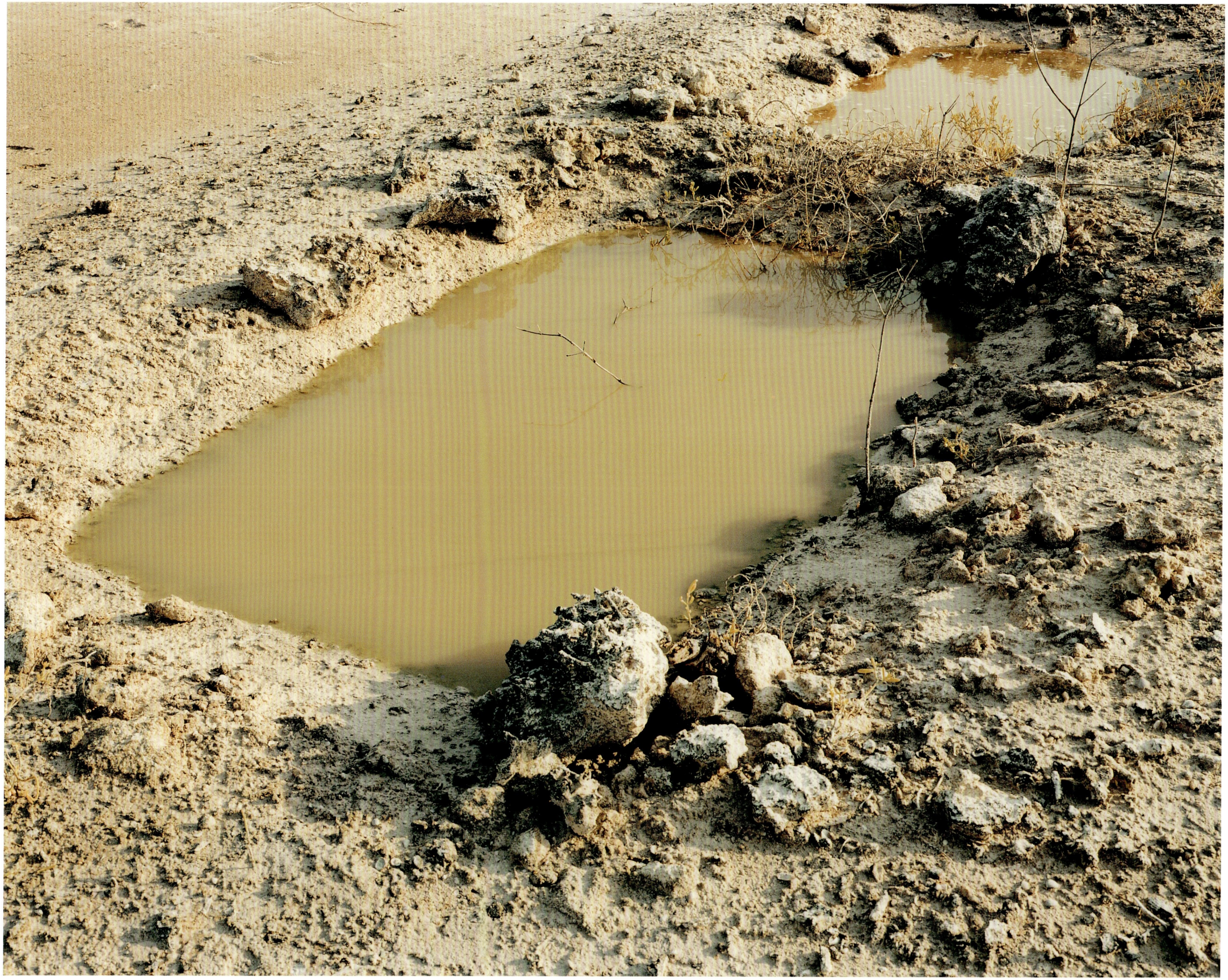

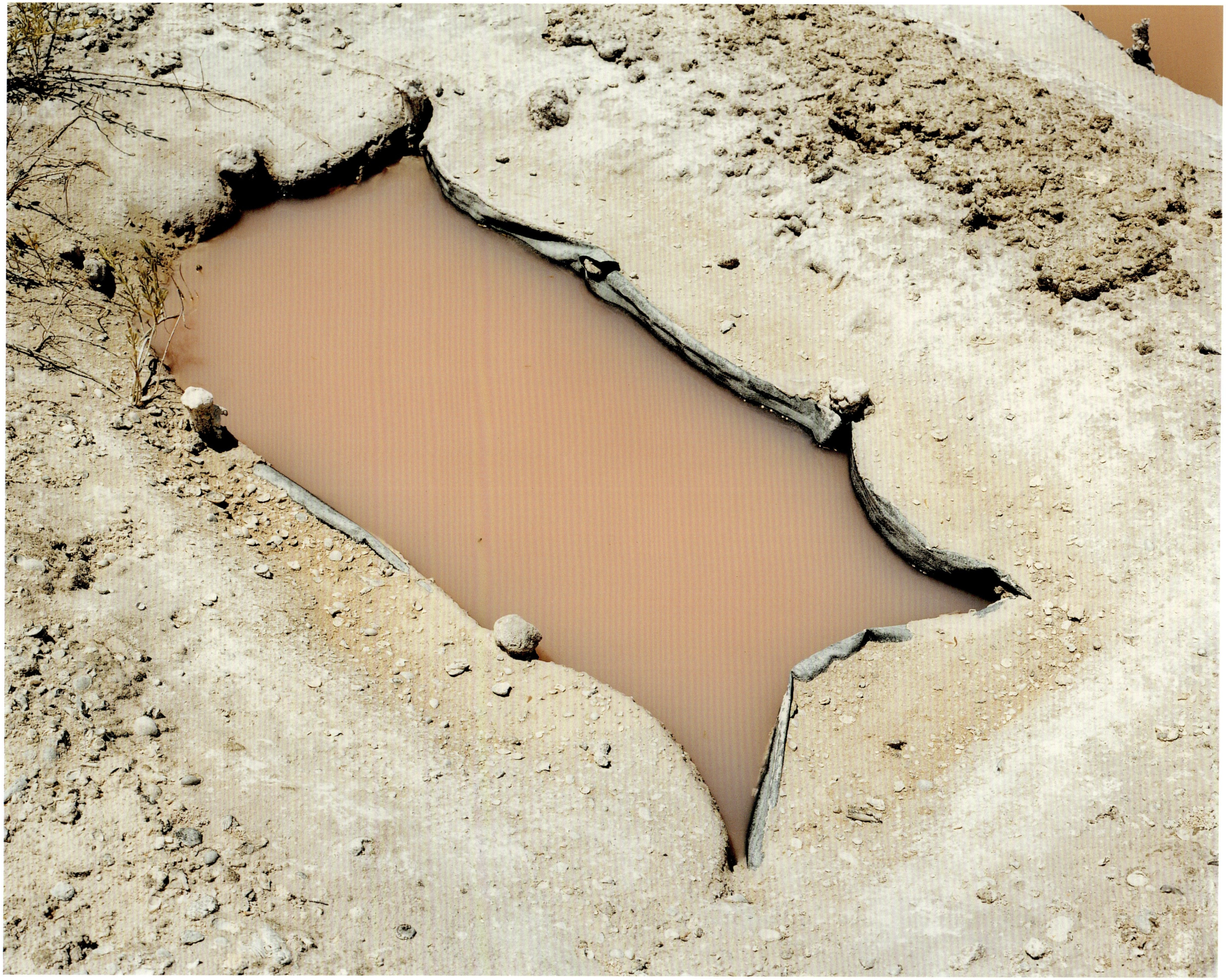

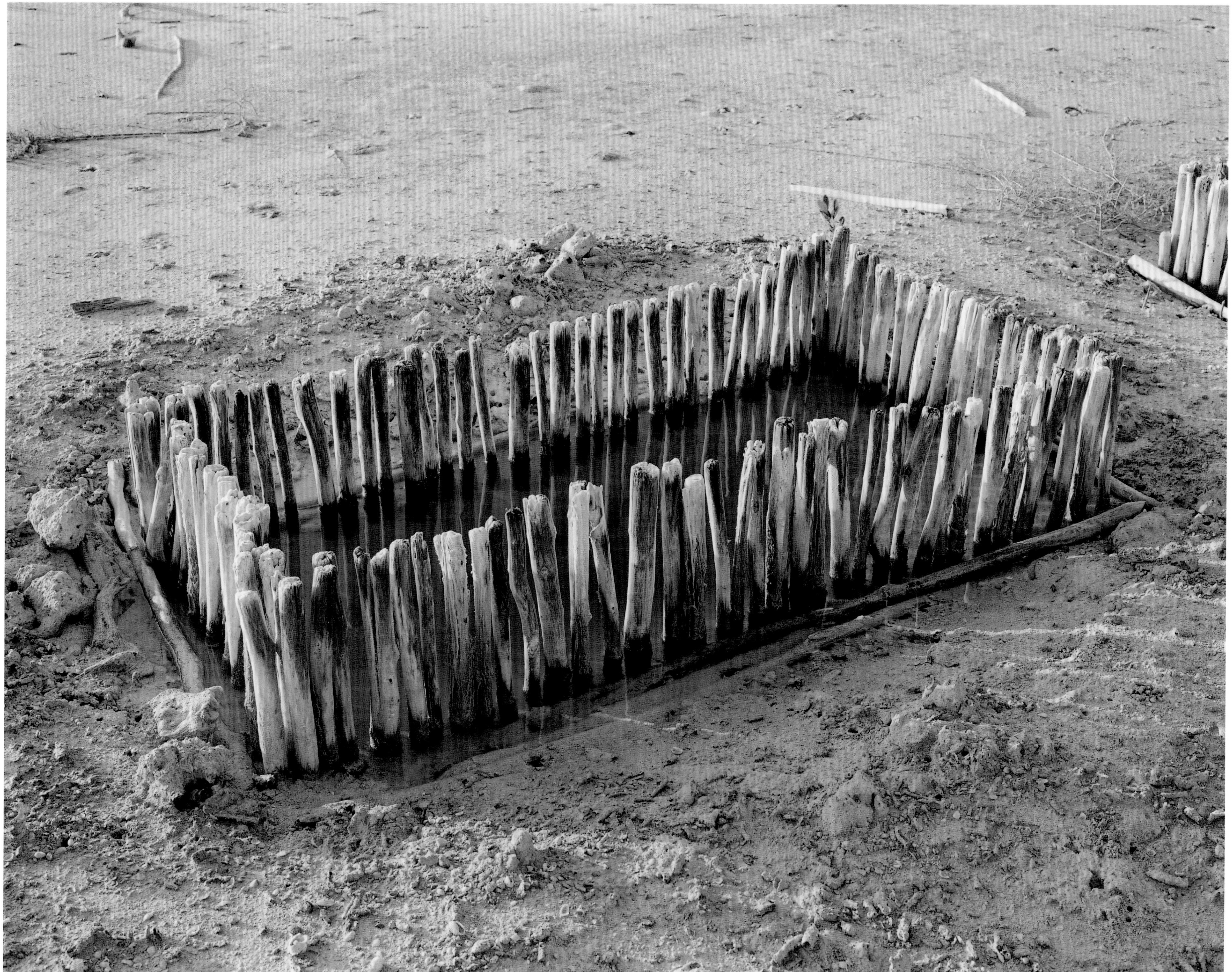

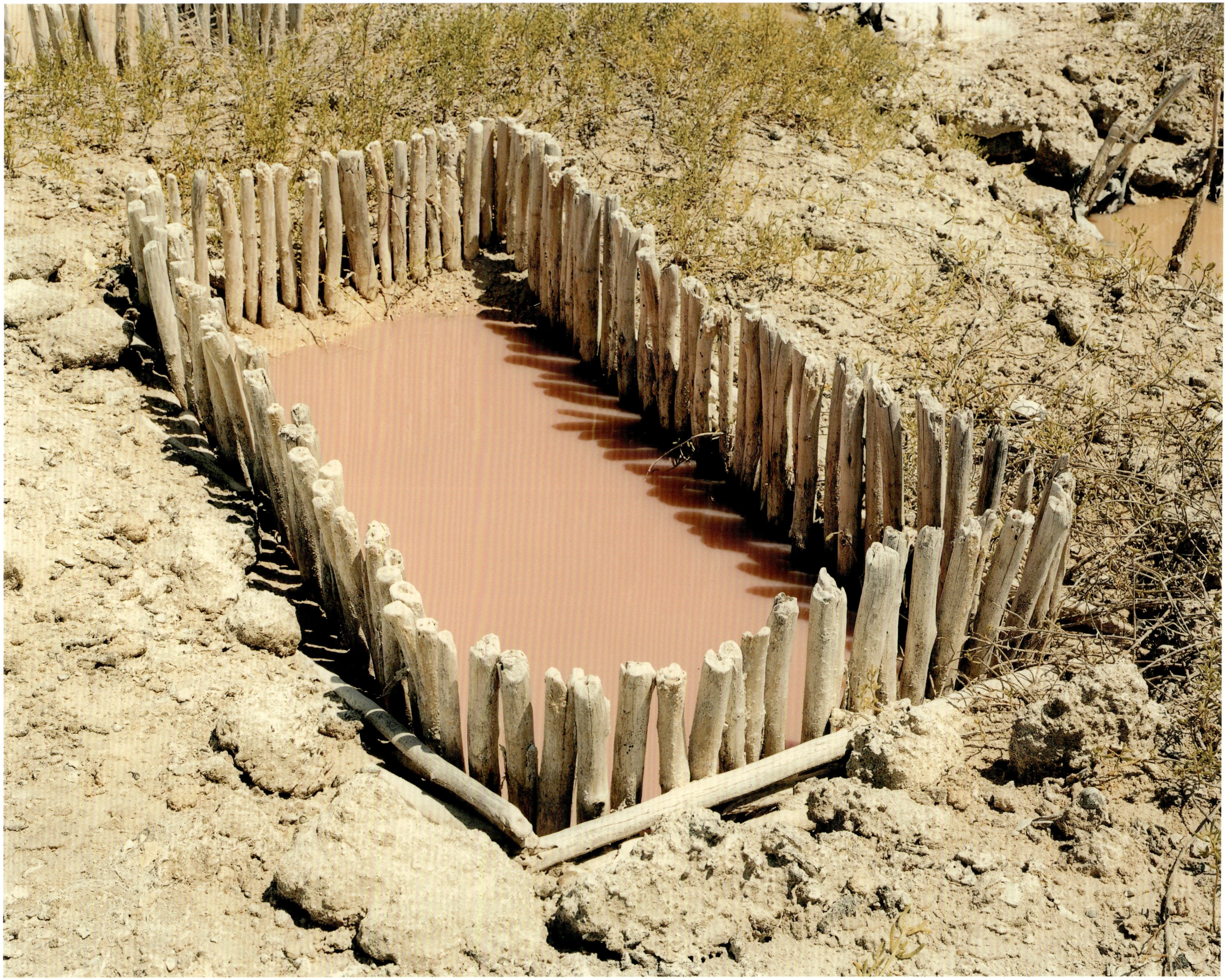

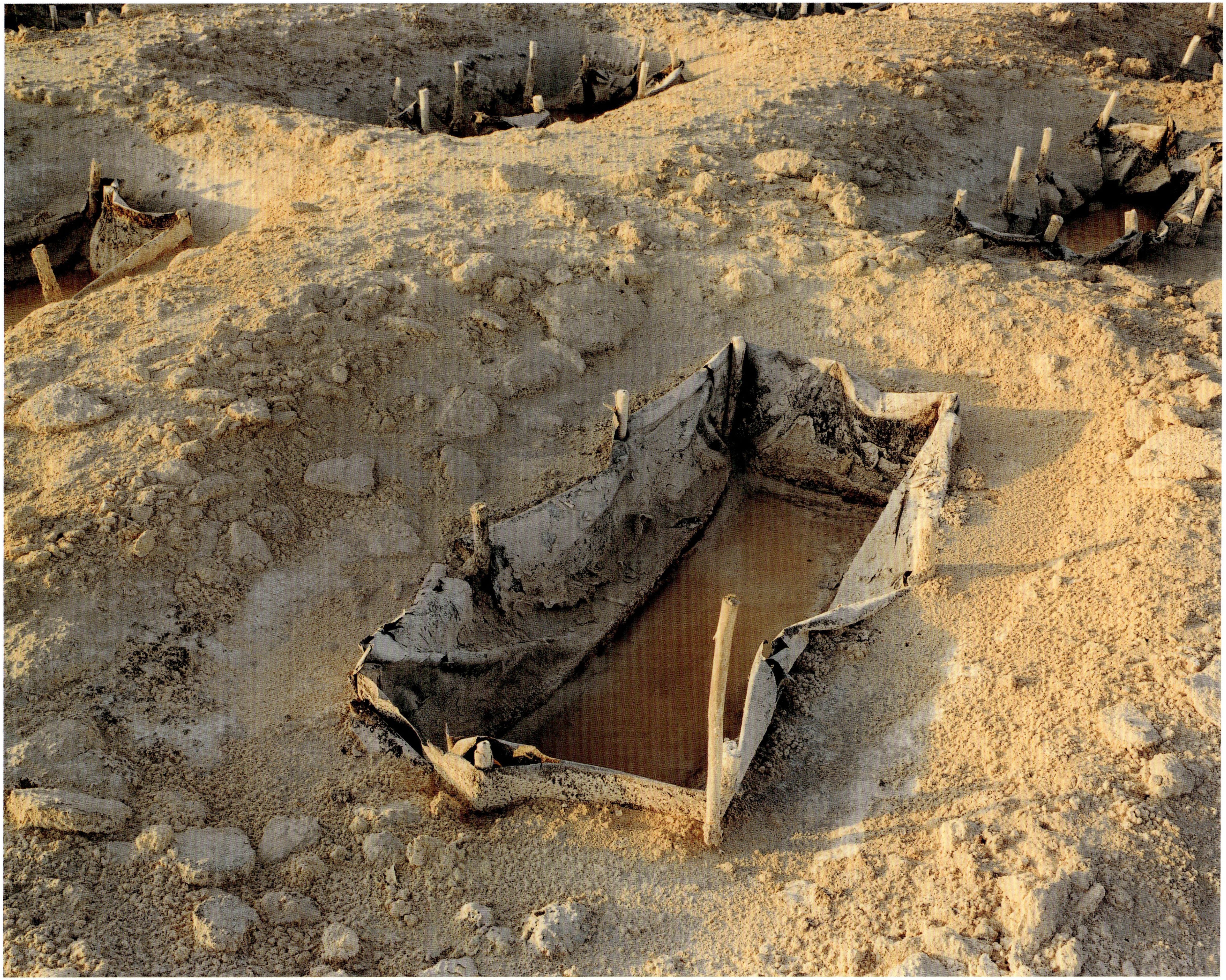

THE EDGE

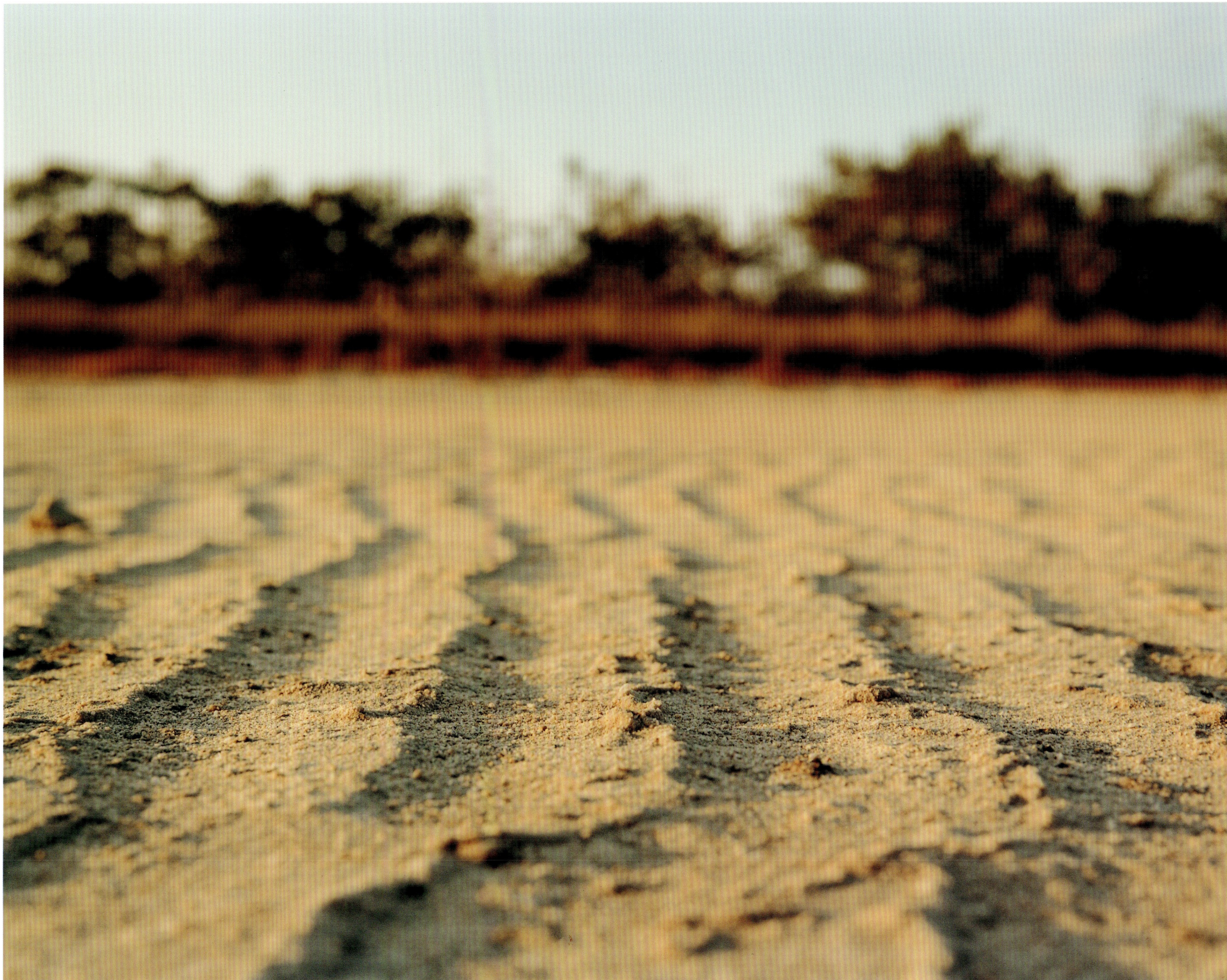

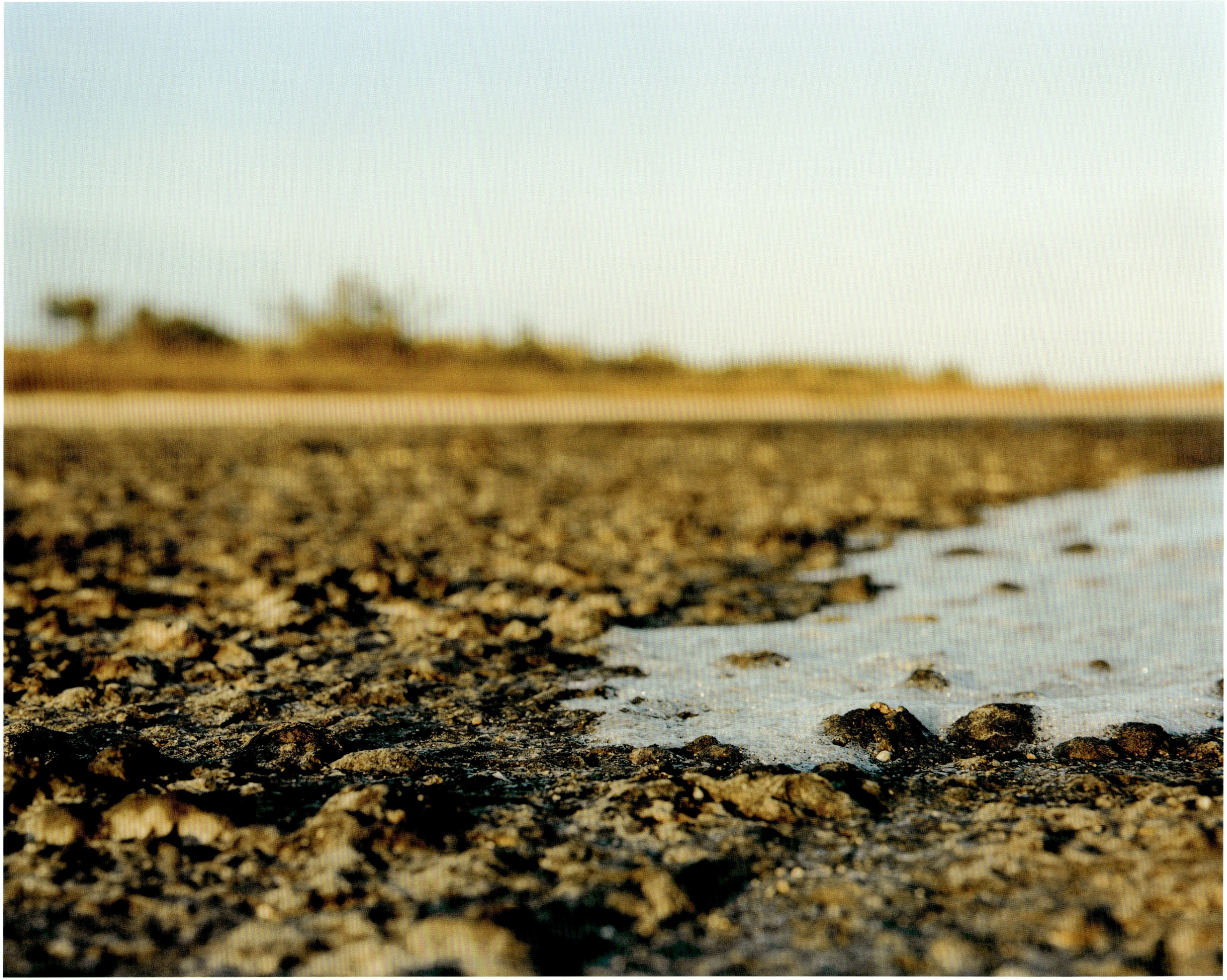

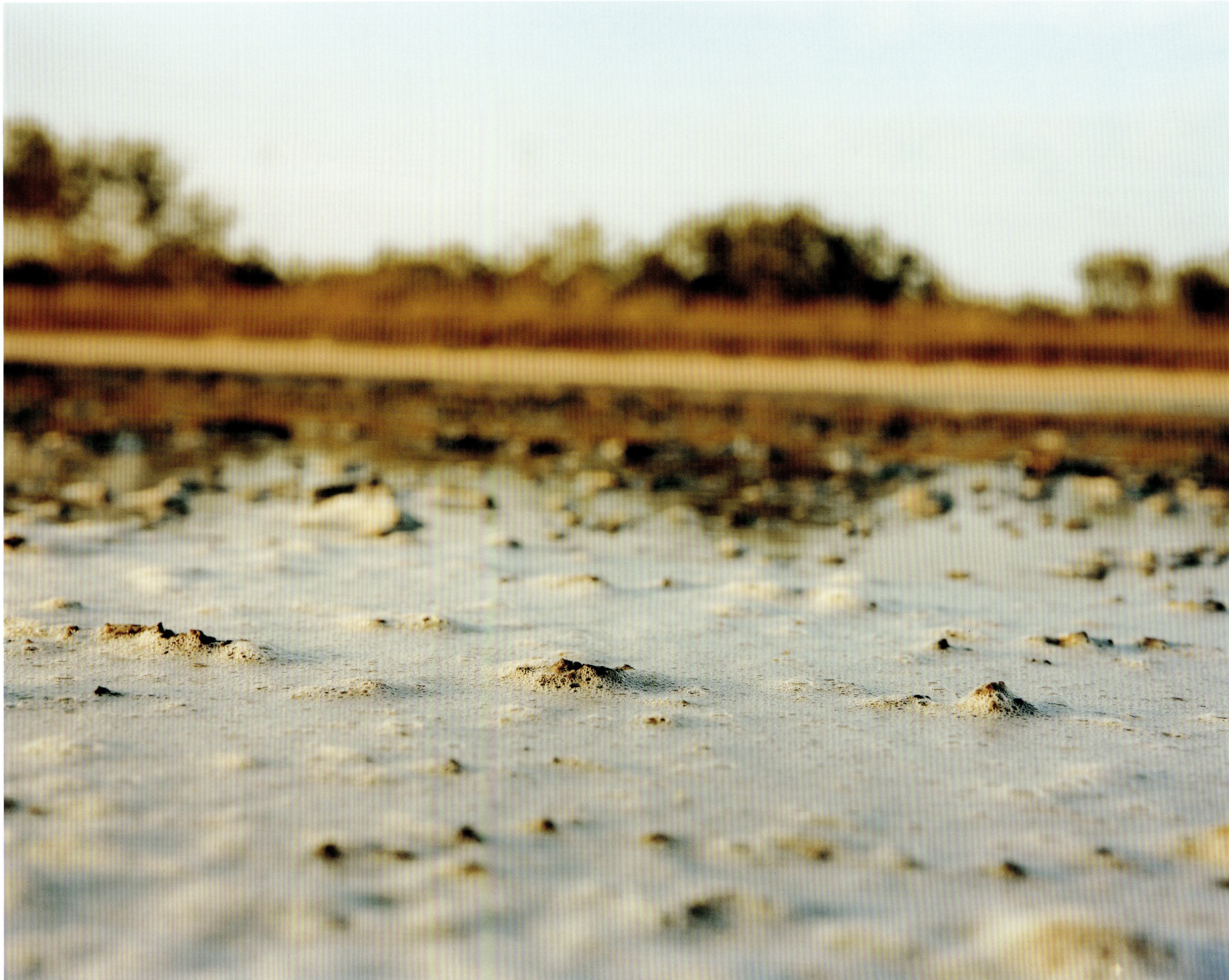

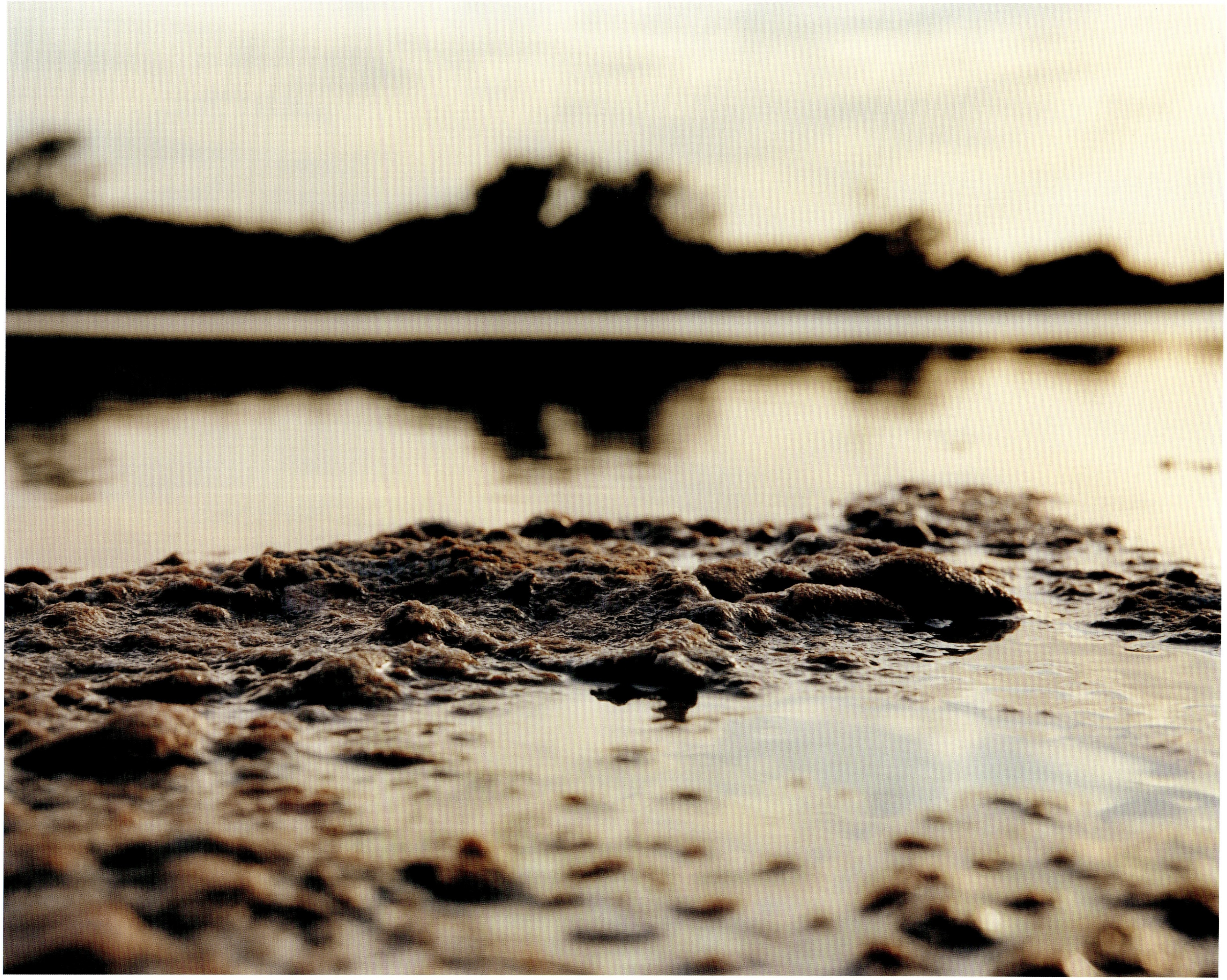

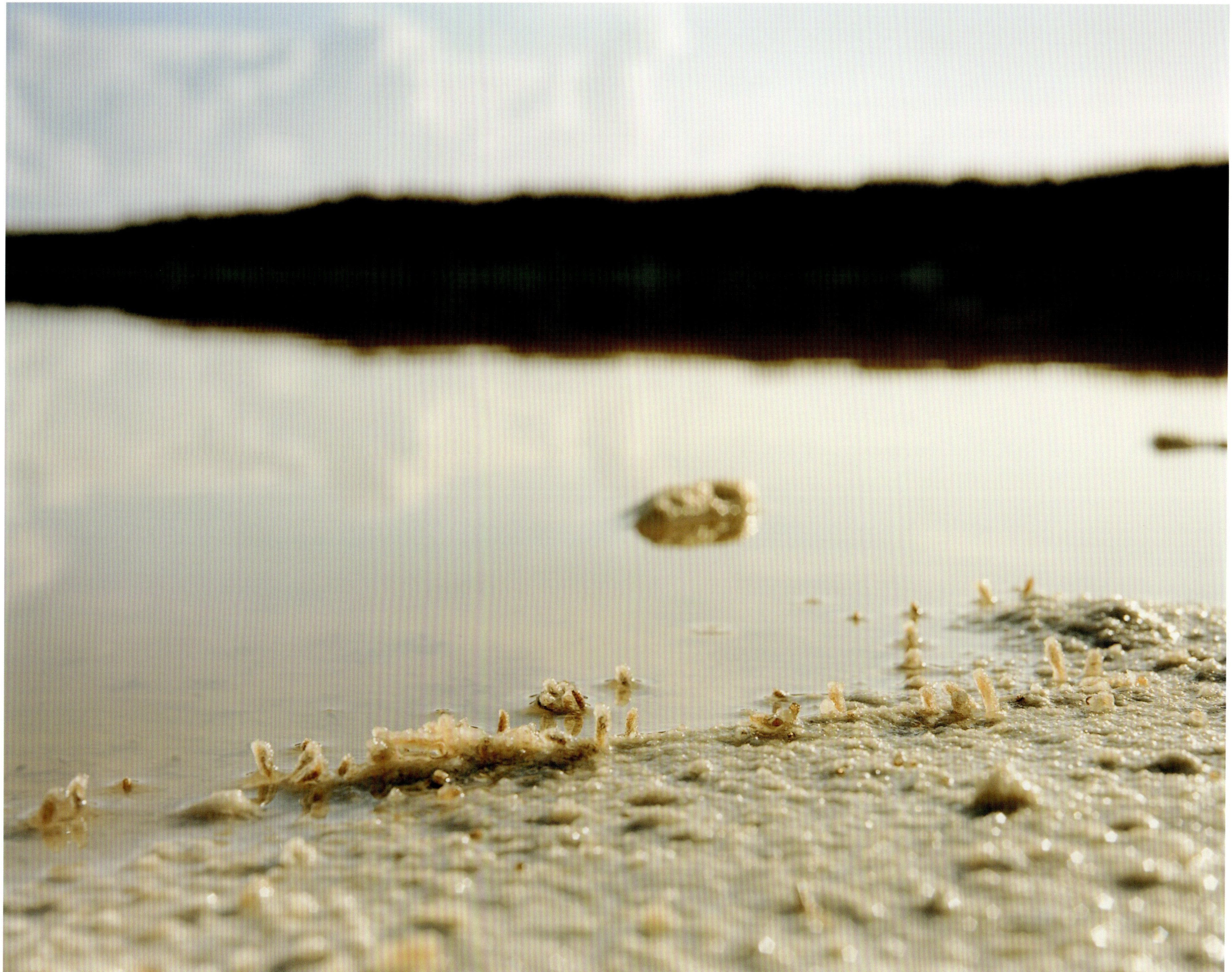

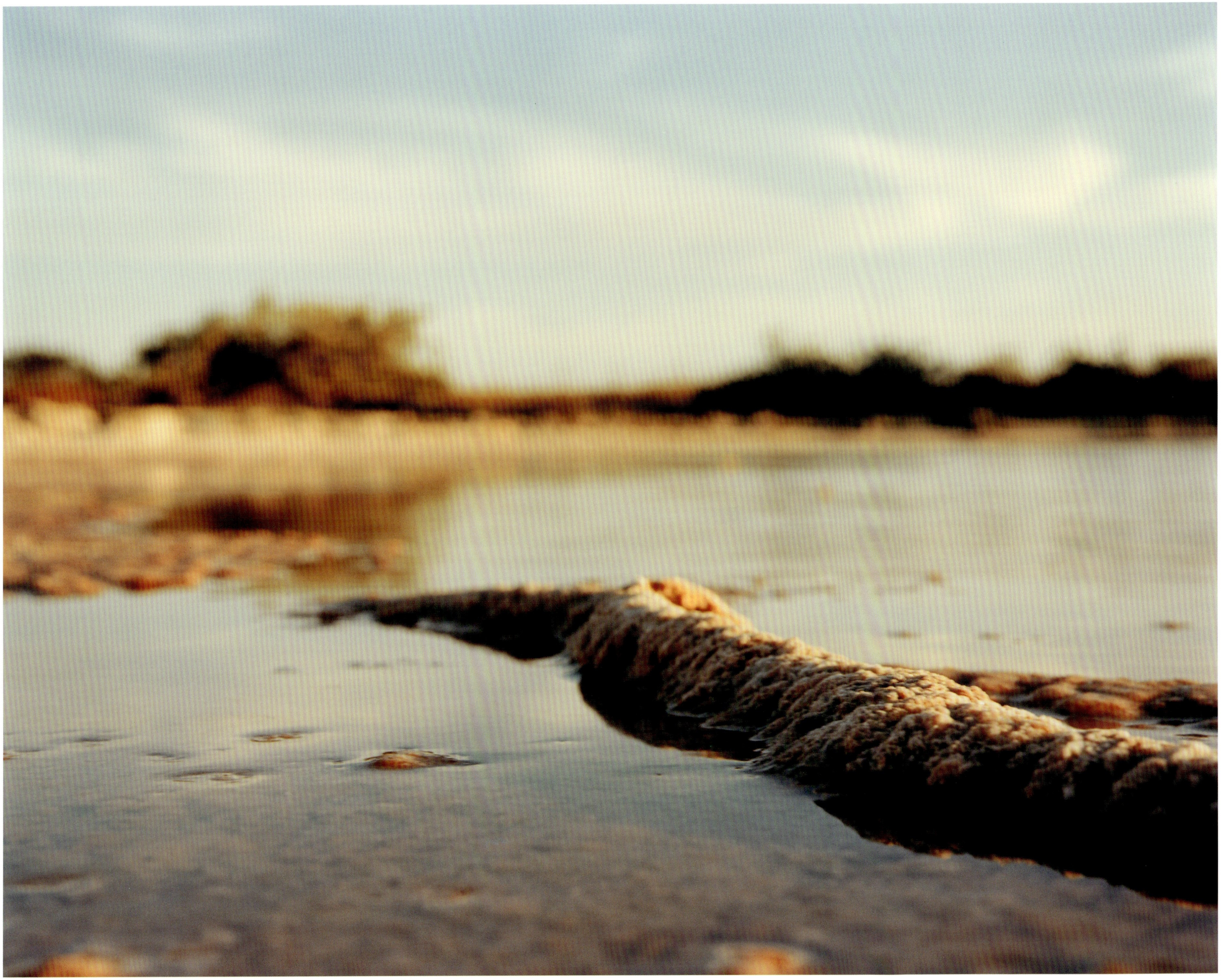

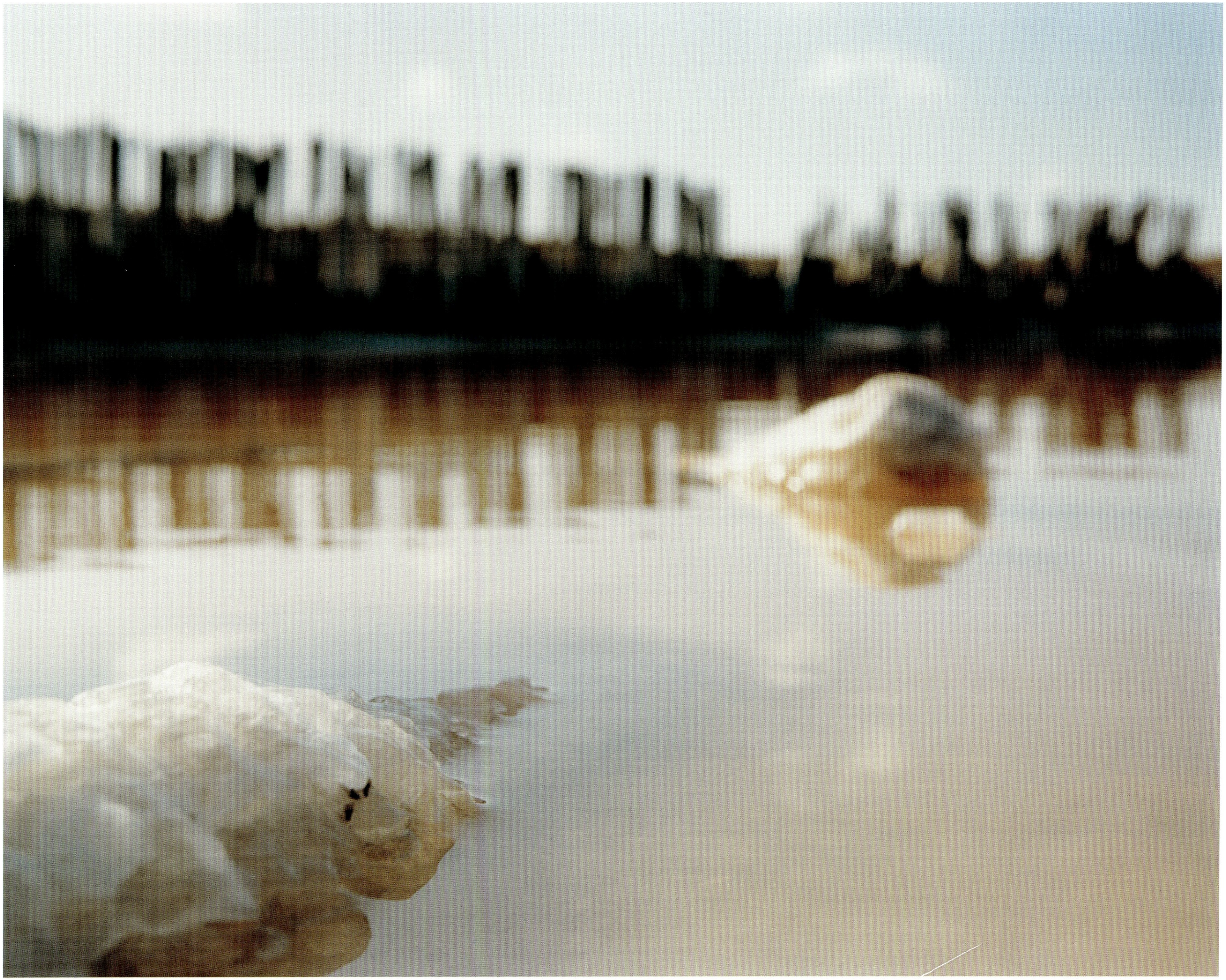

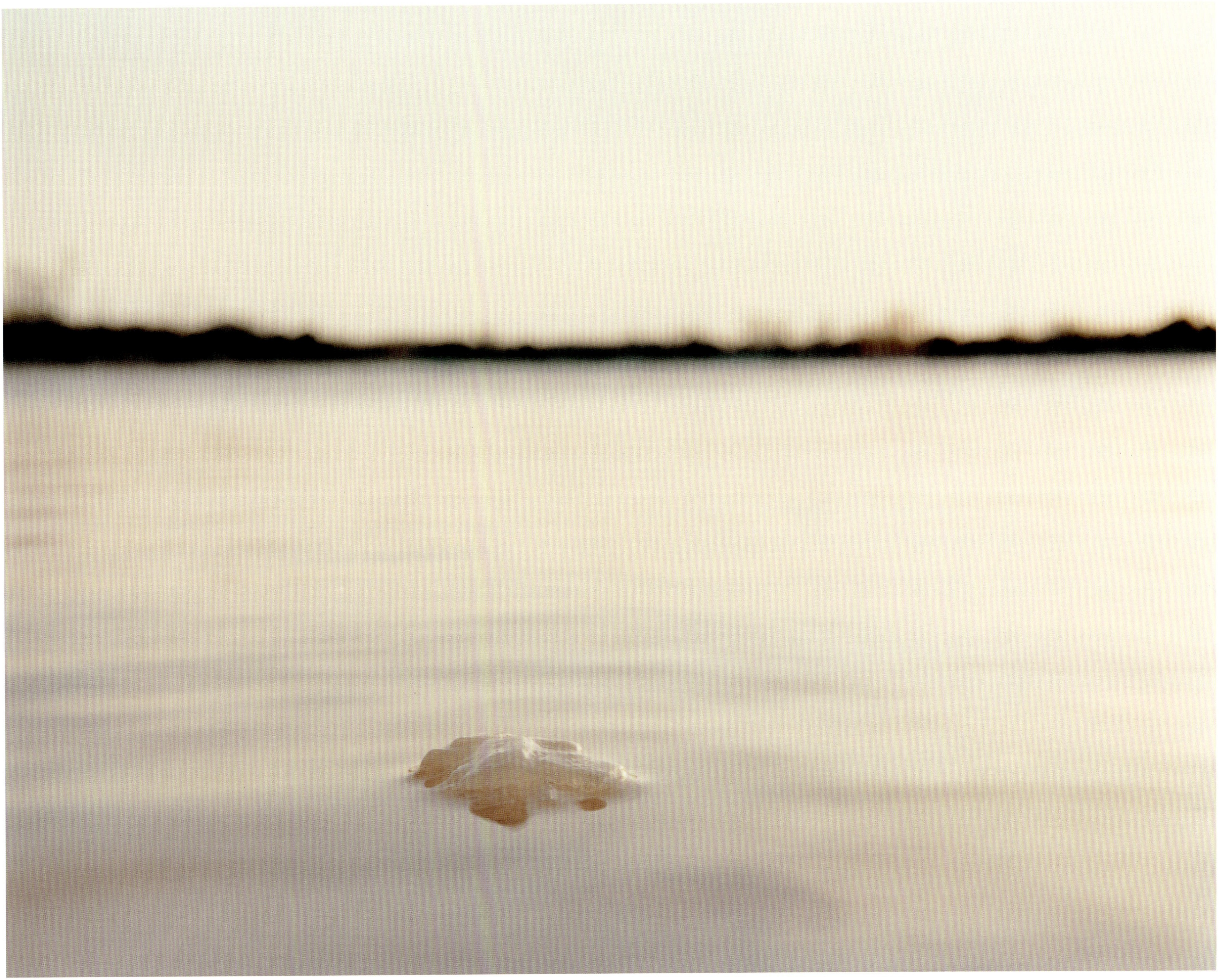

THE LAGOON

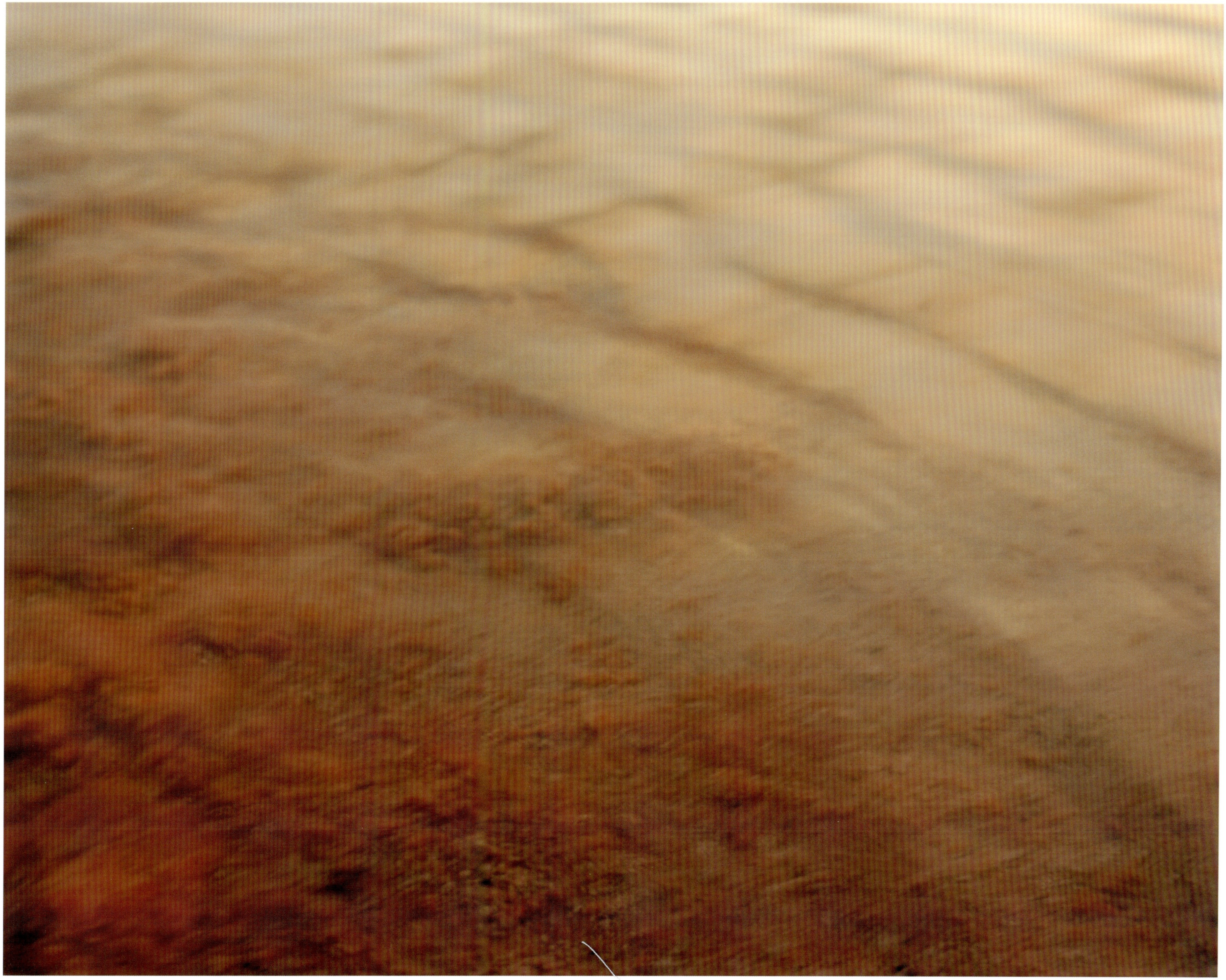

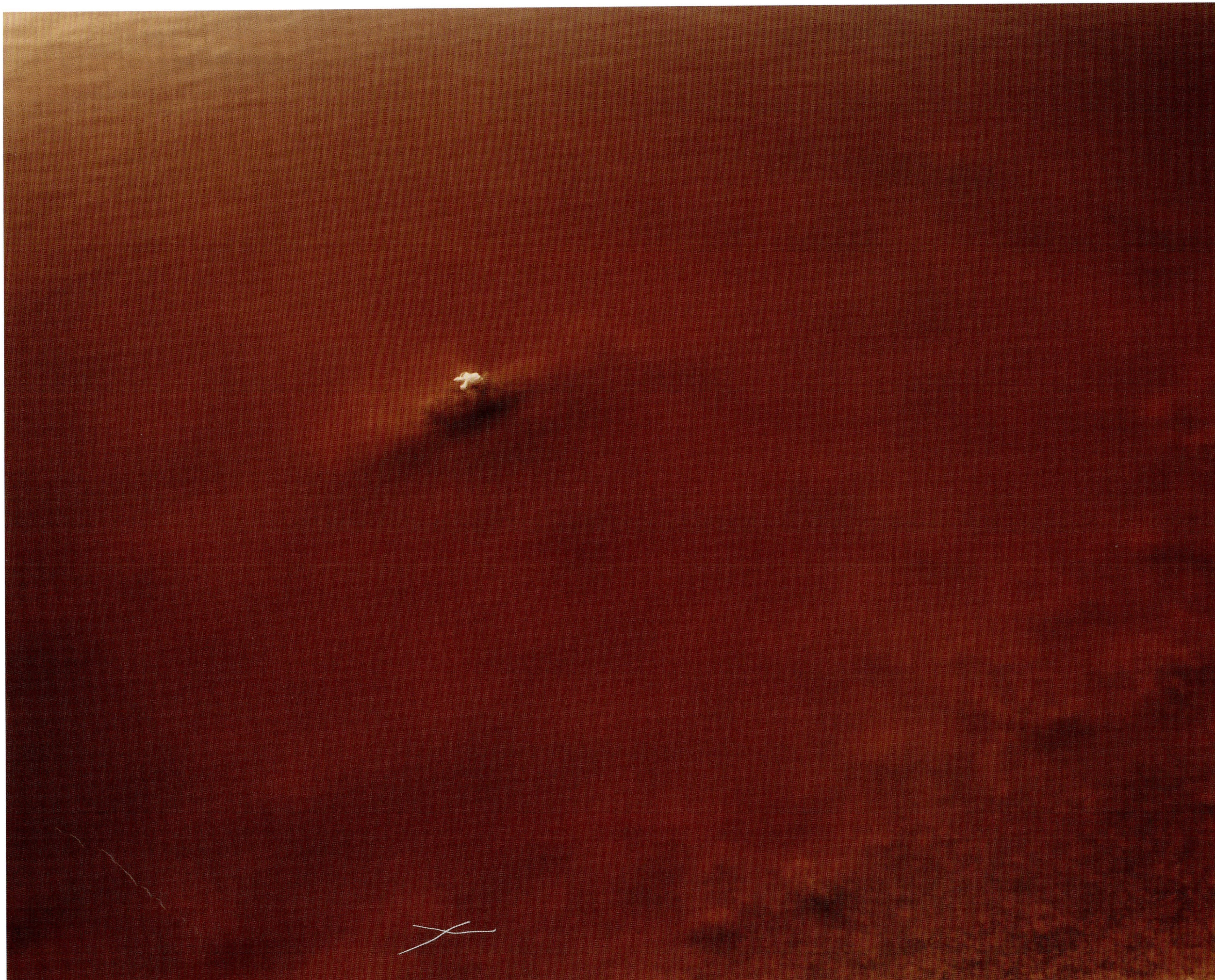

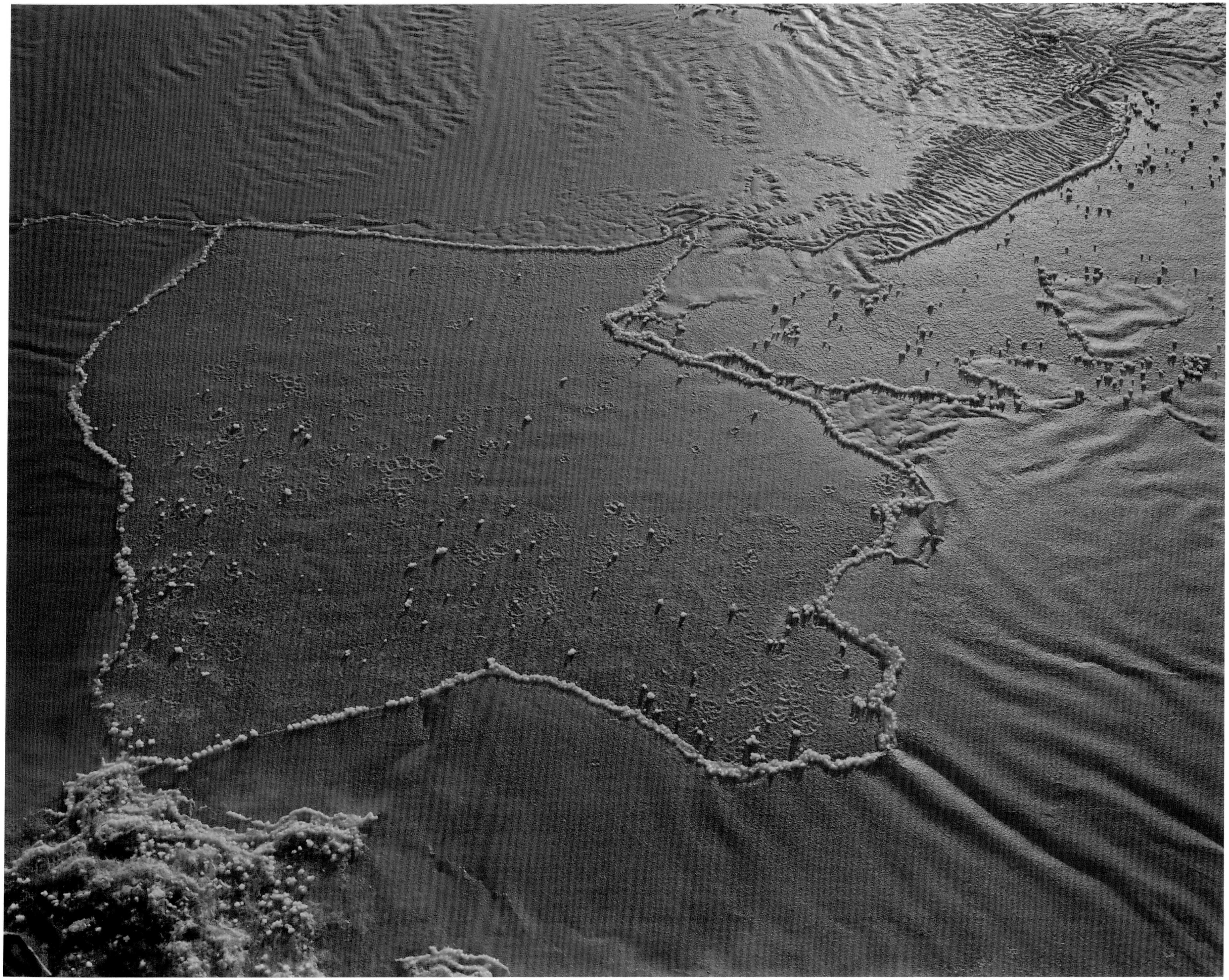

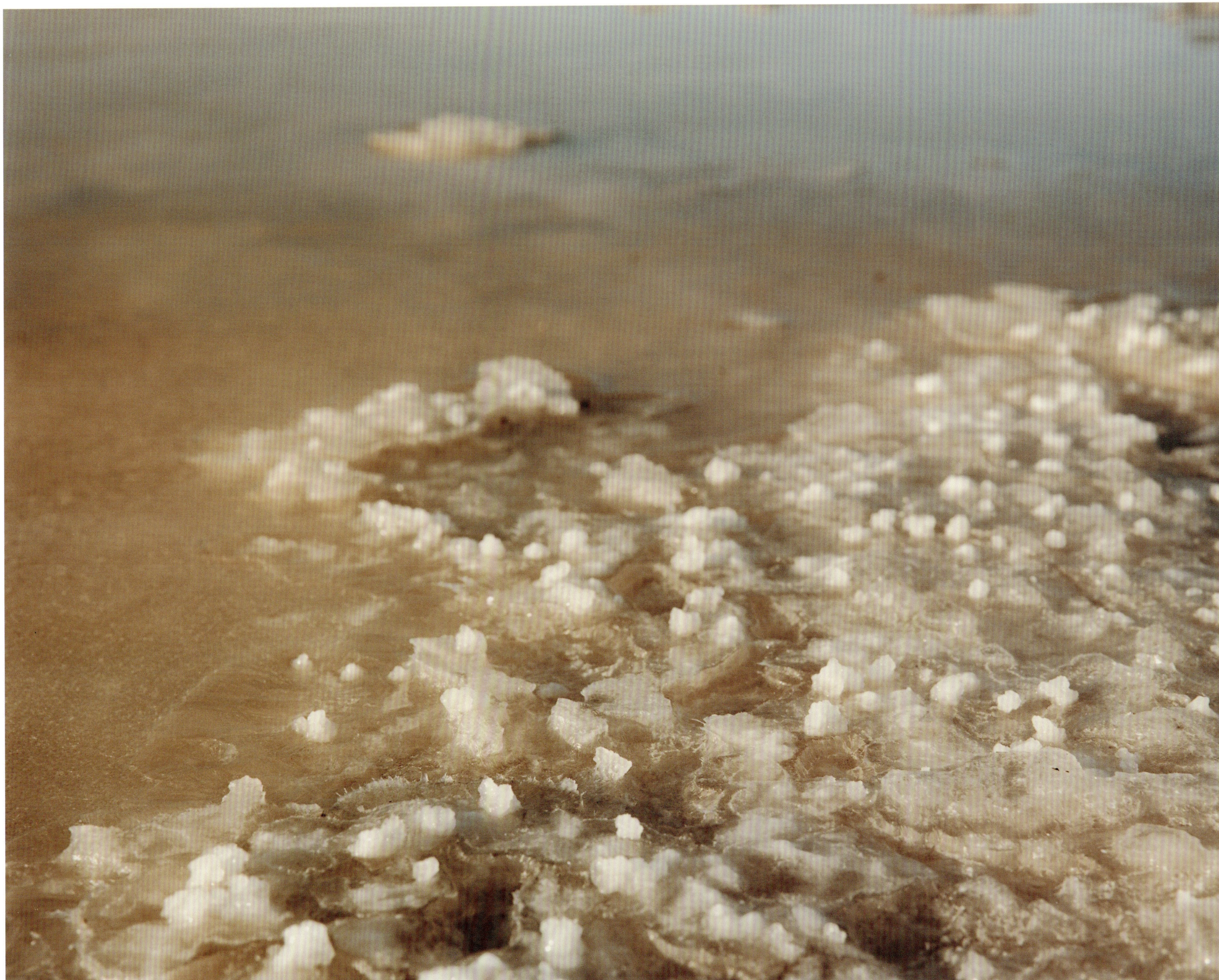

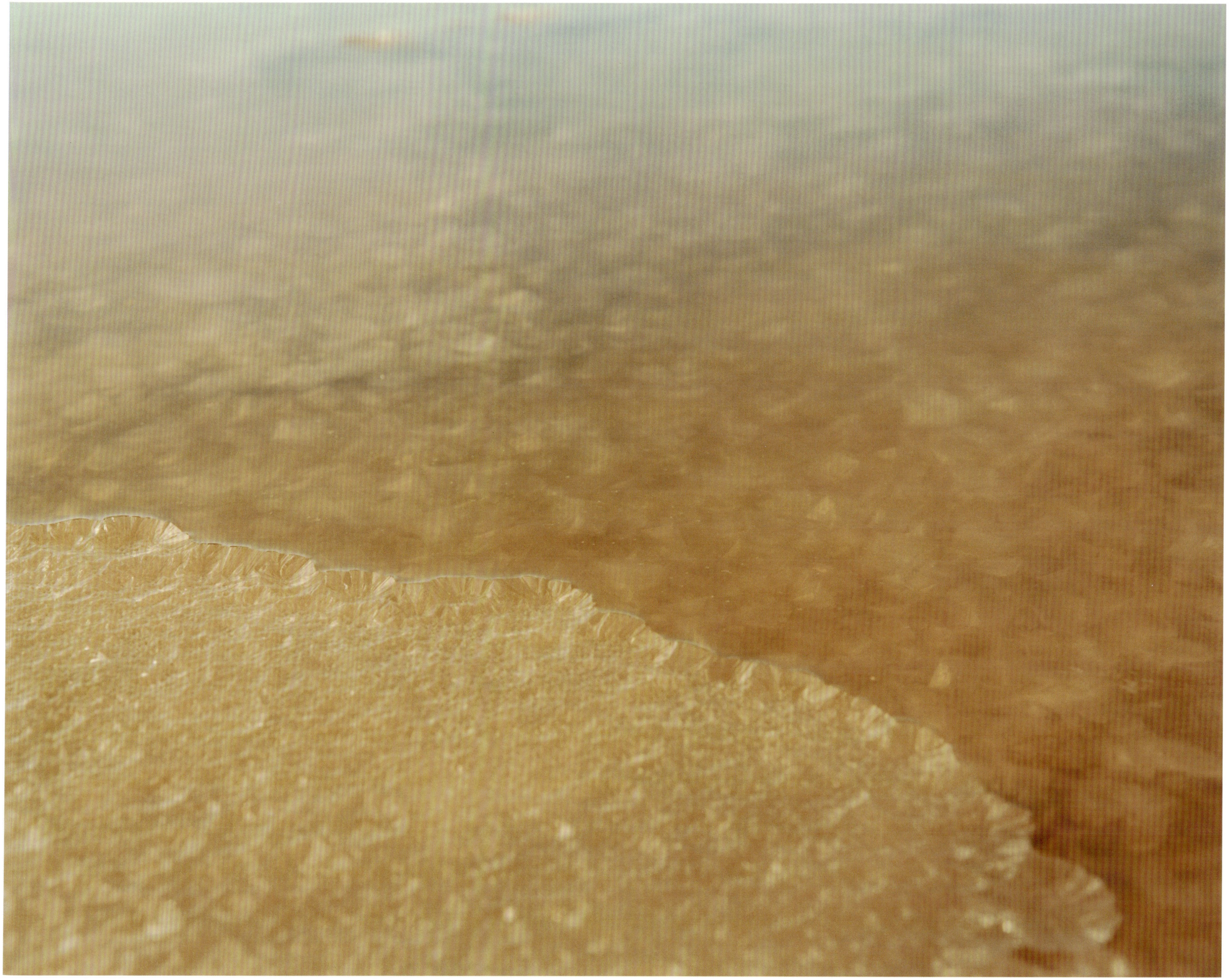

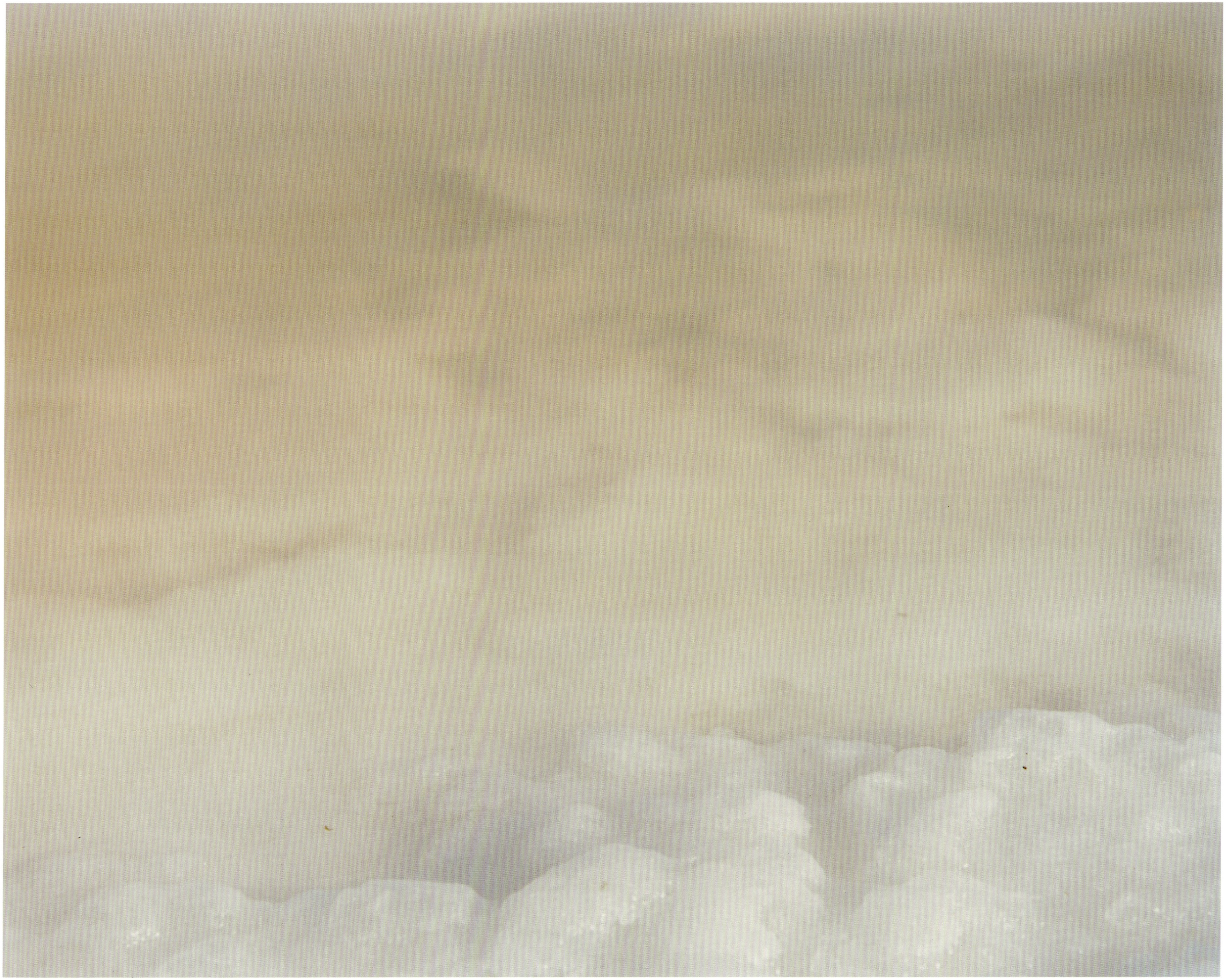

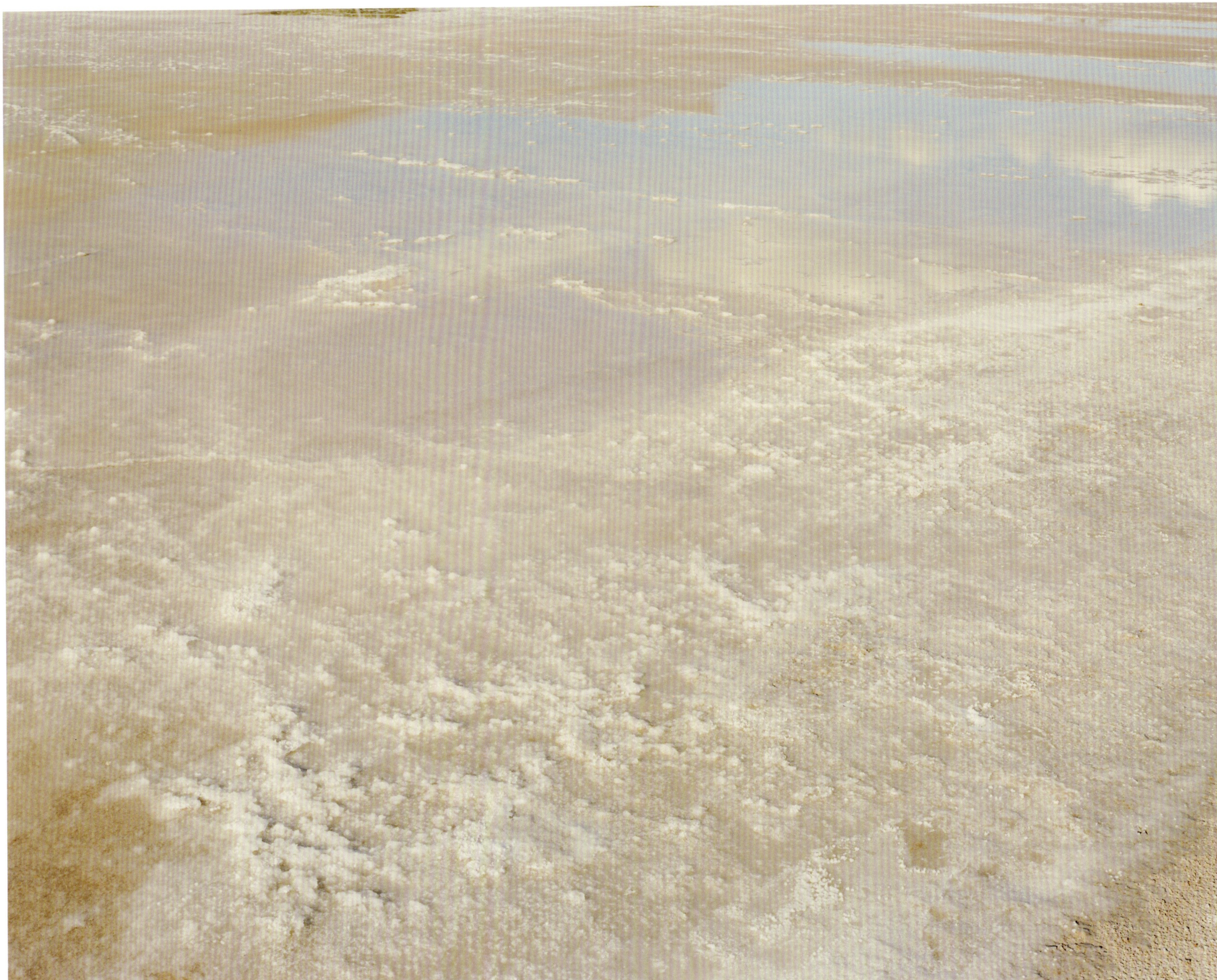

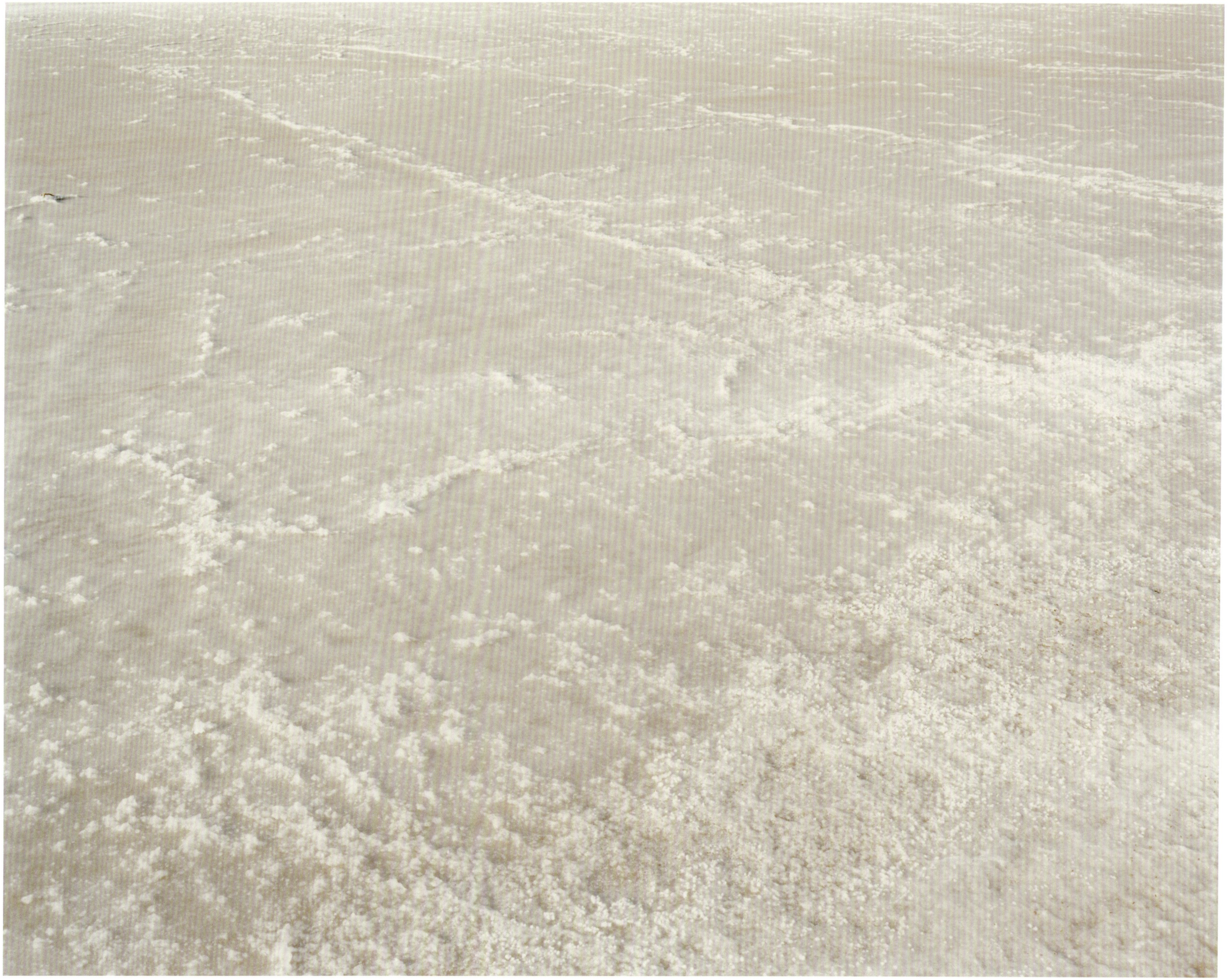

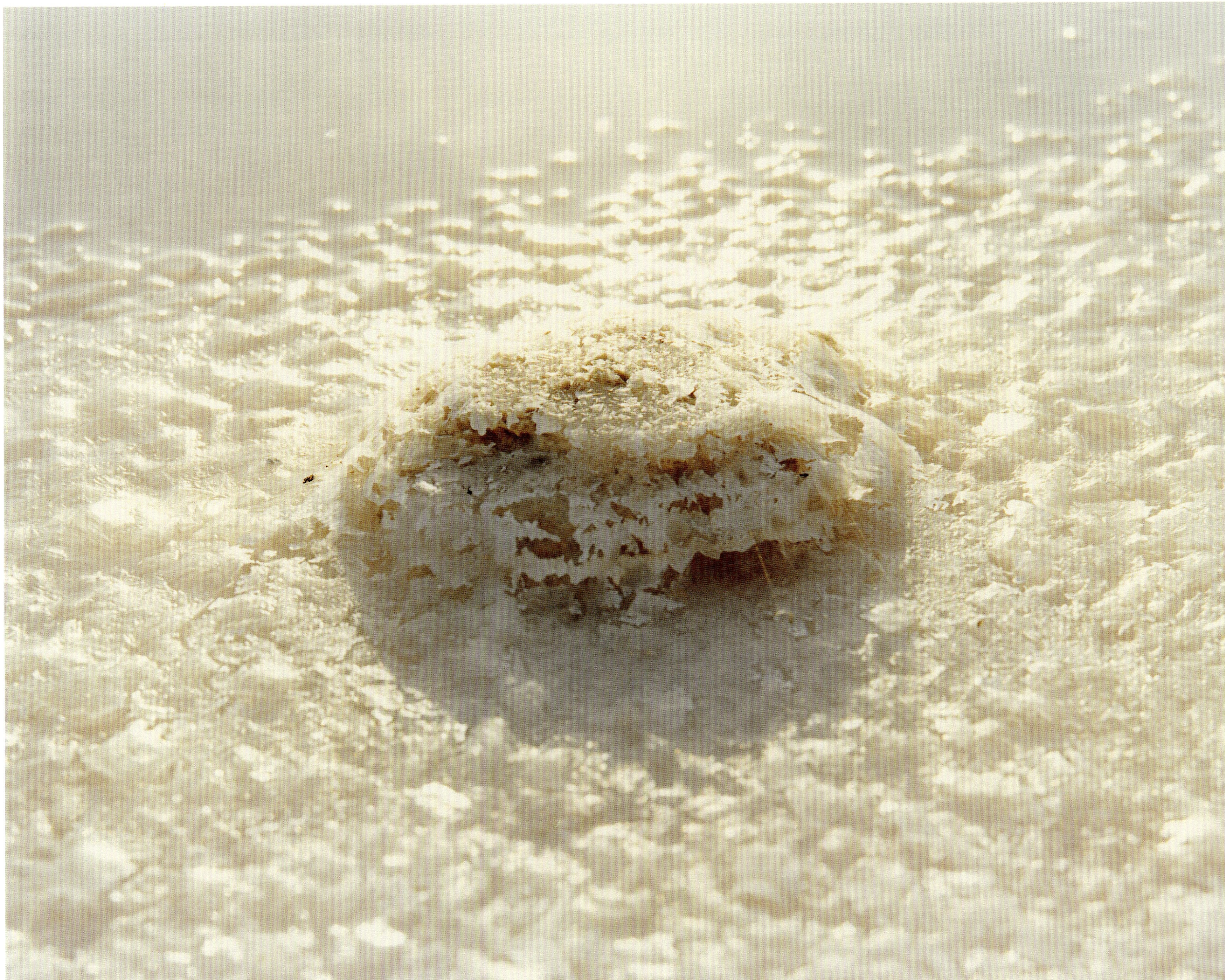

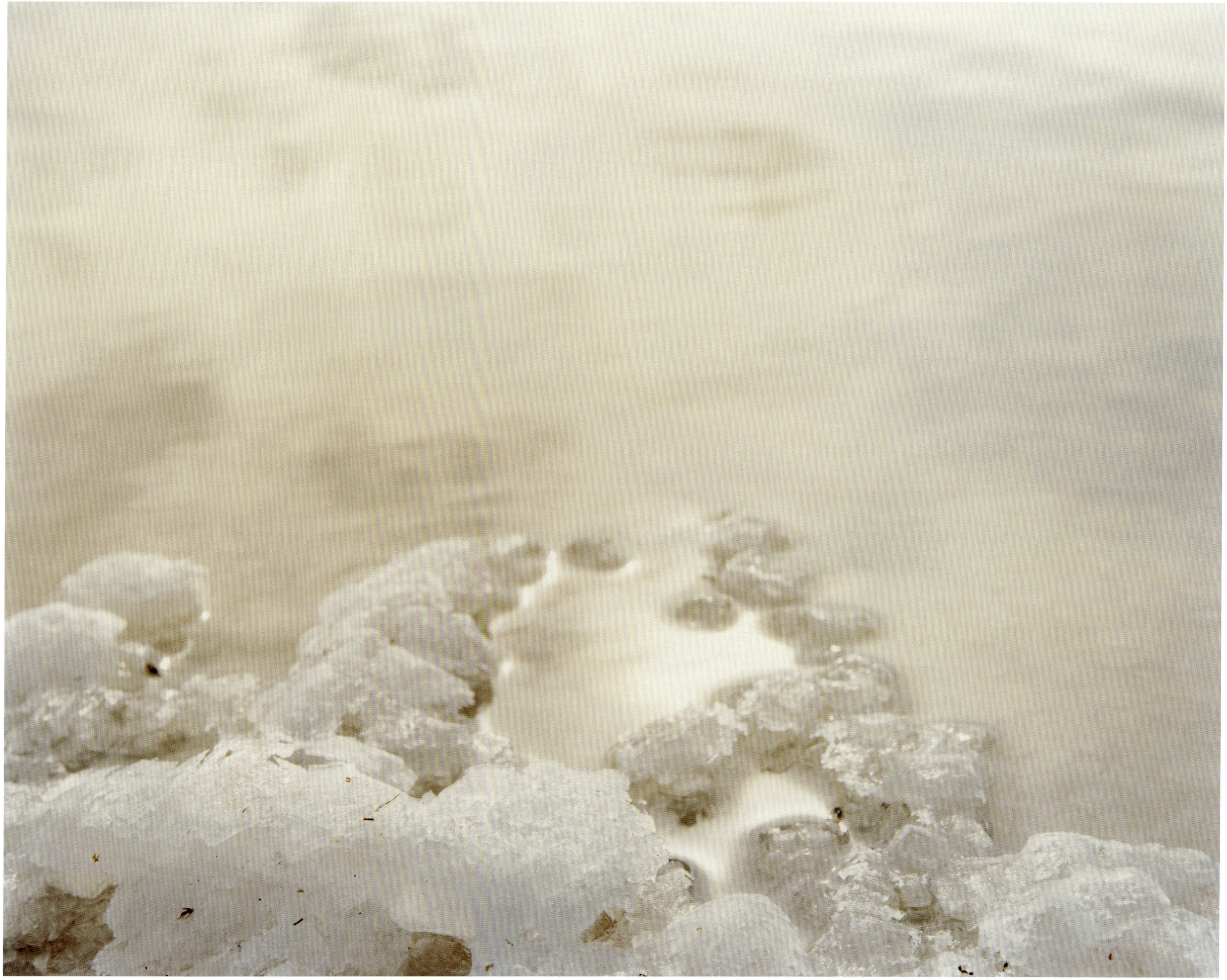

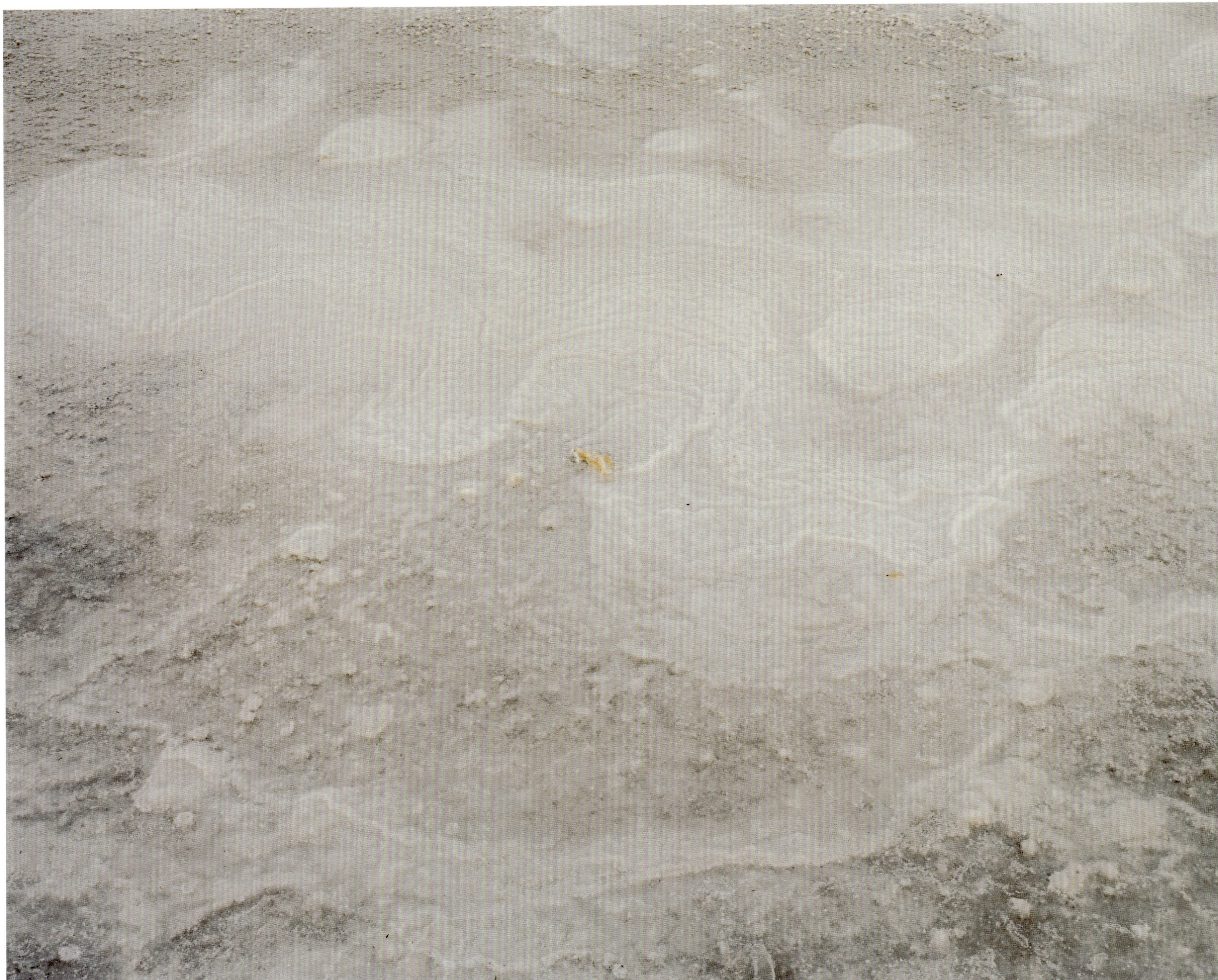

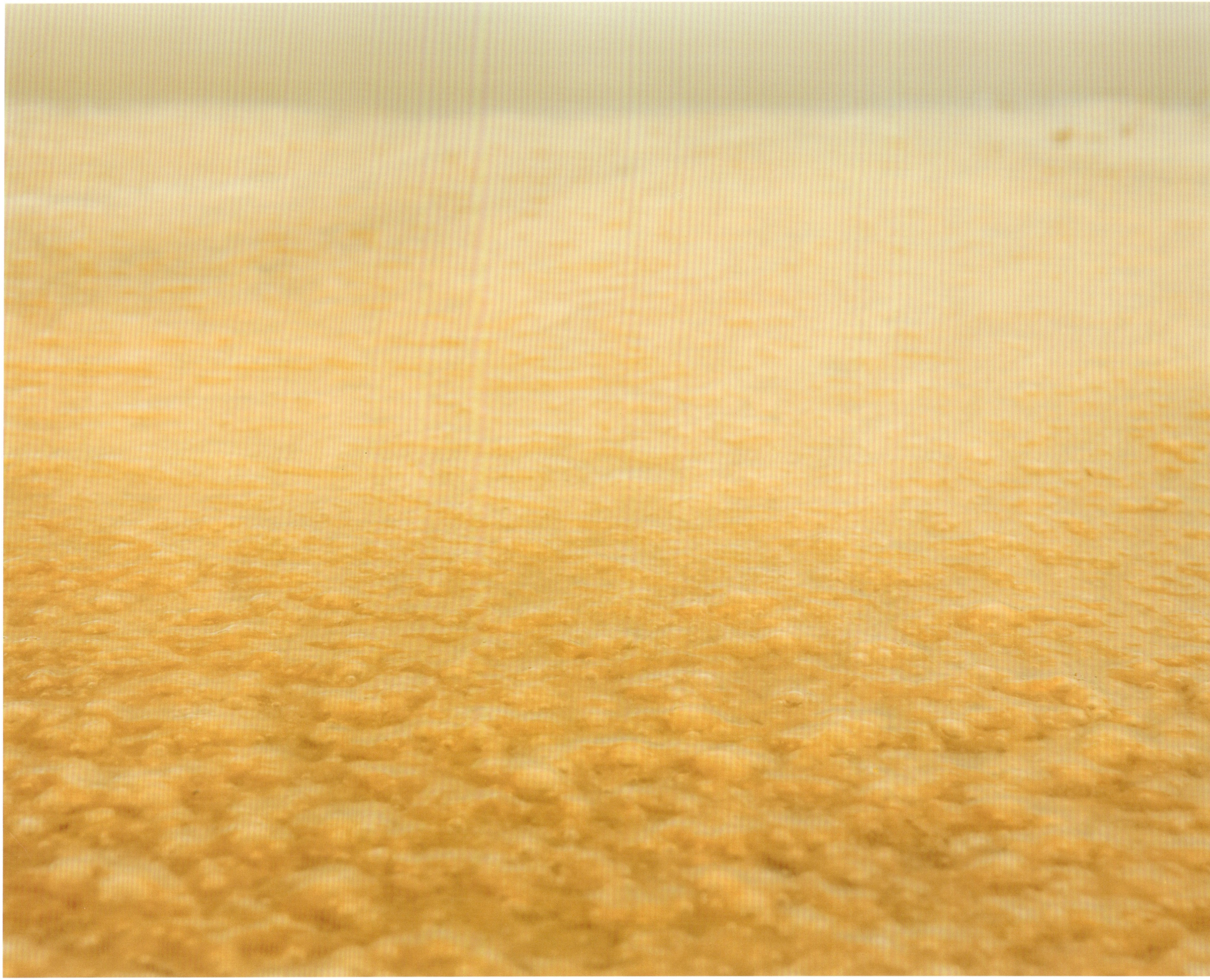

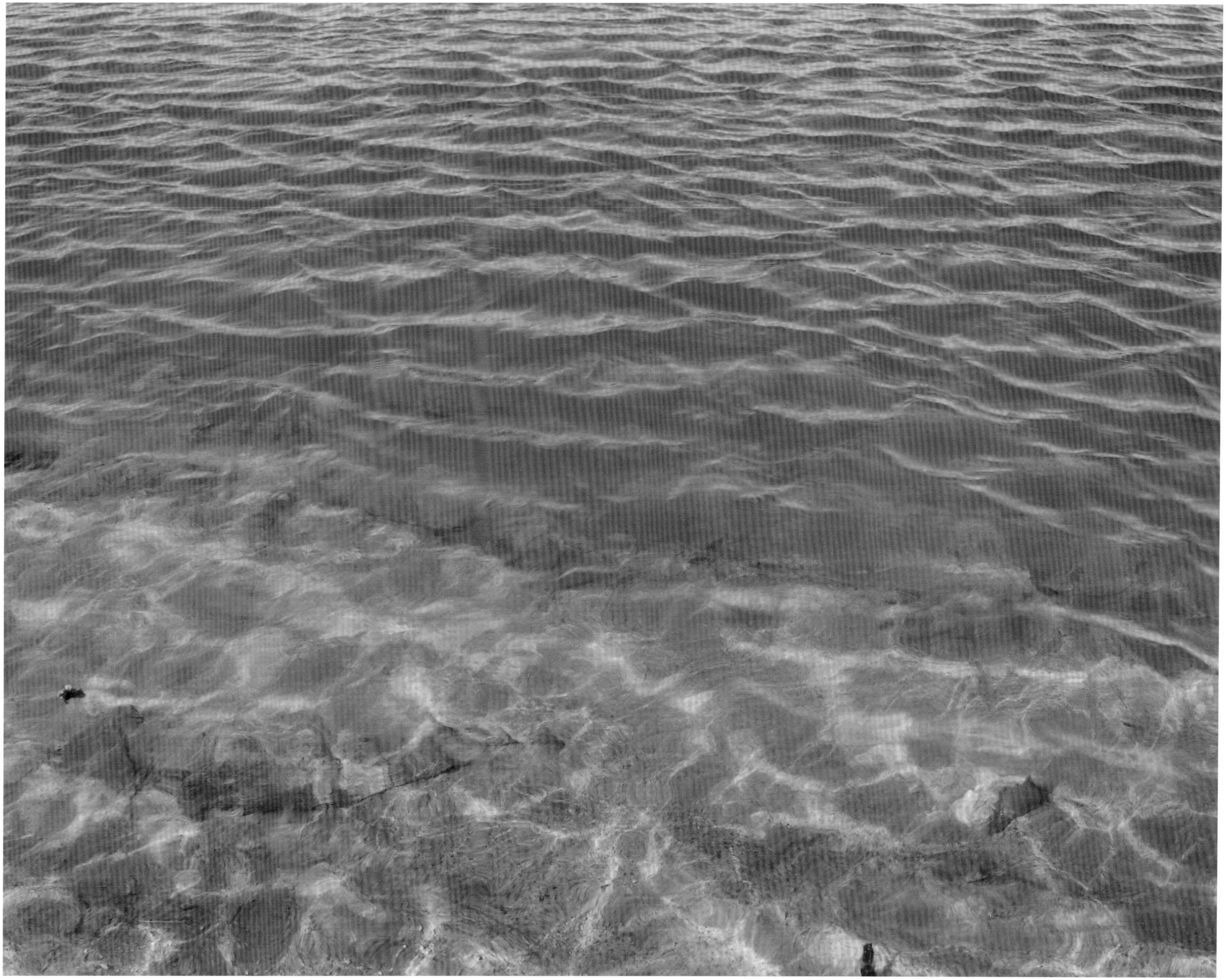

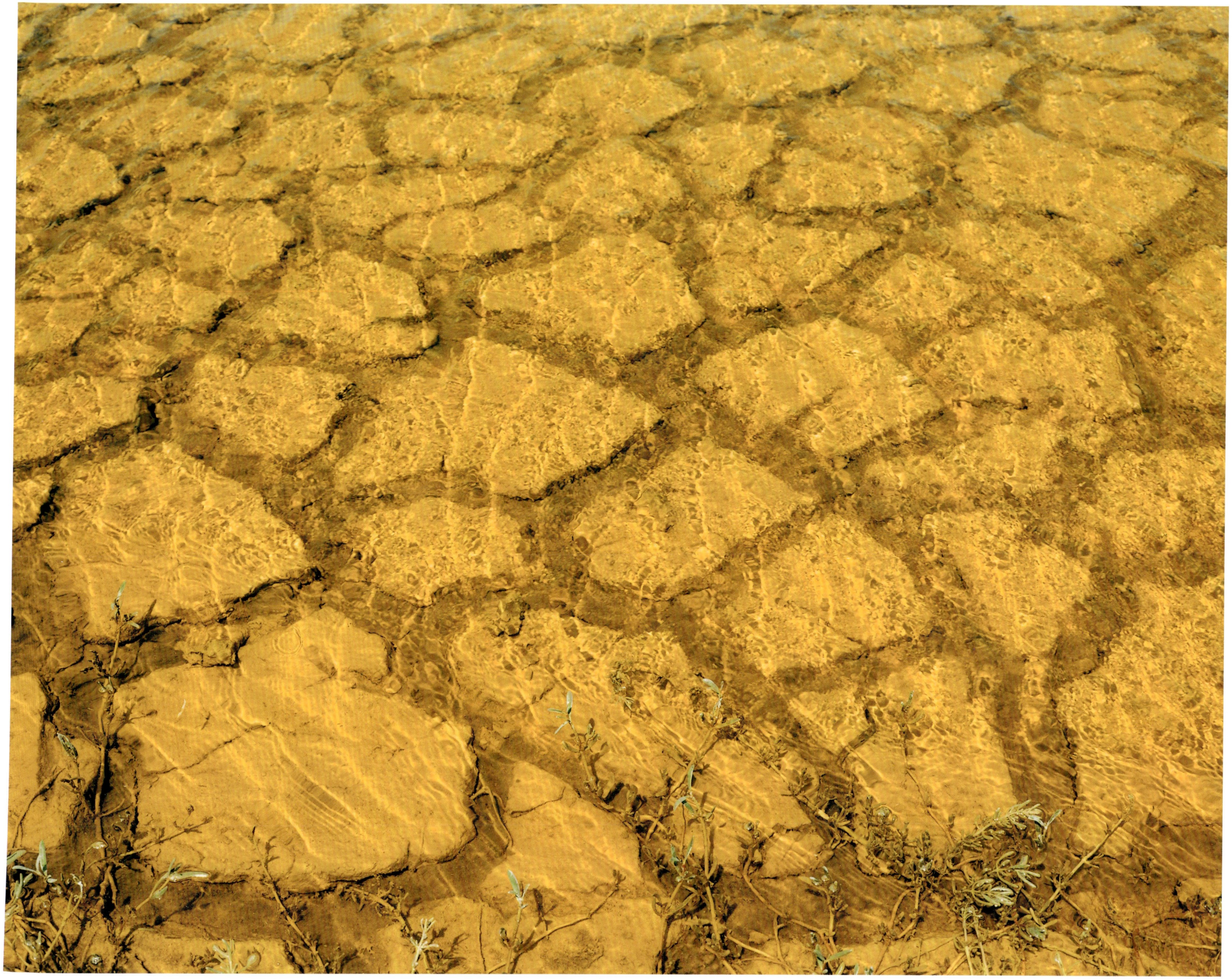

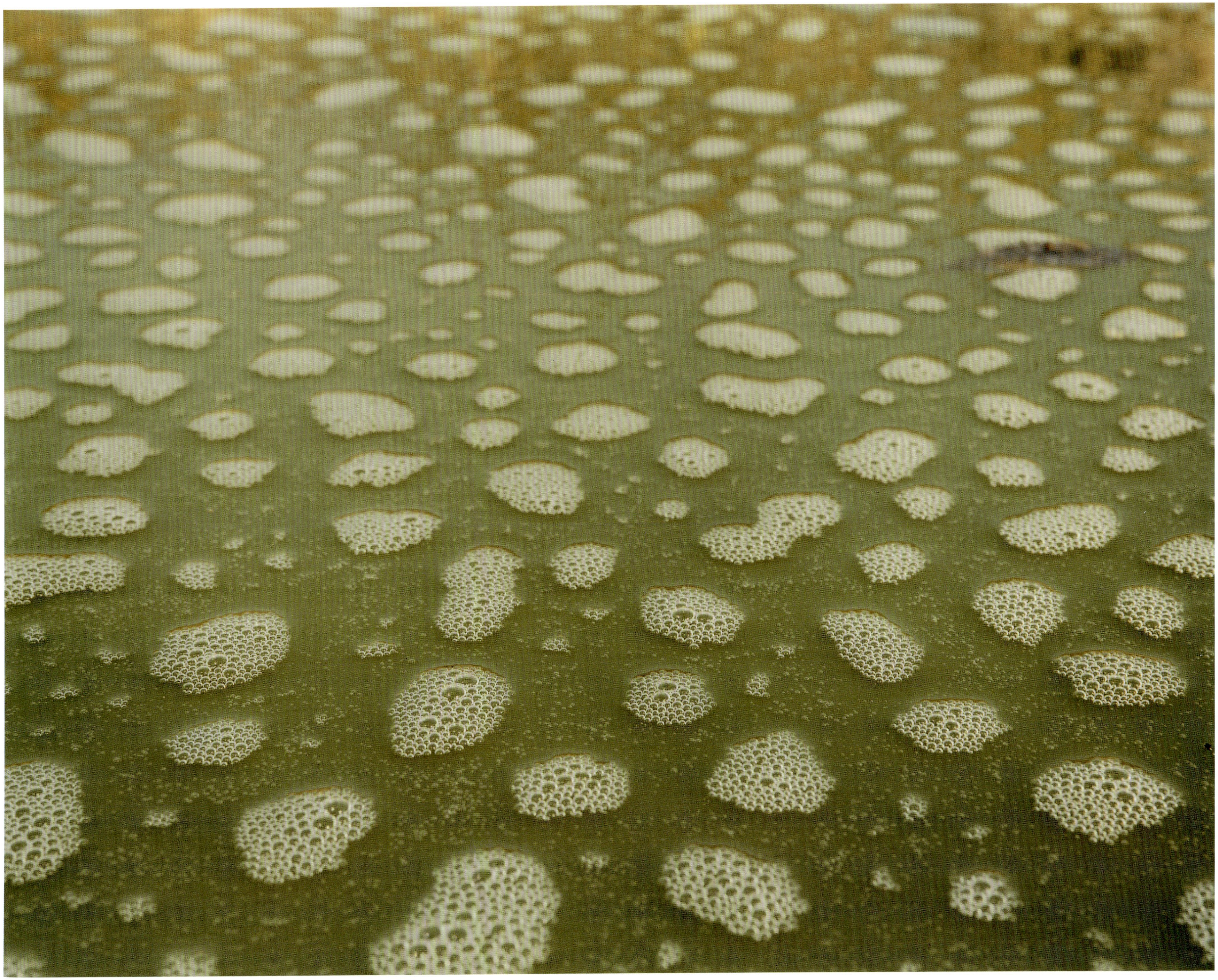

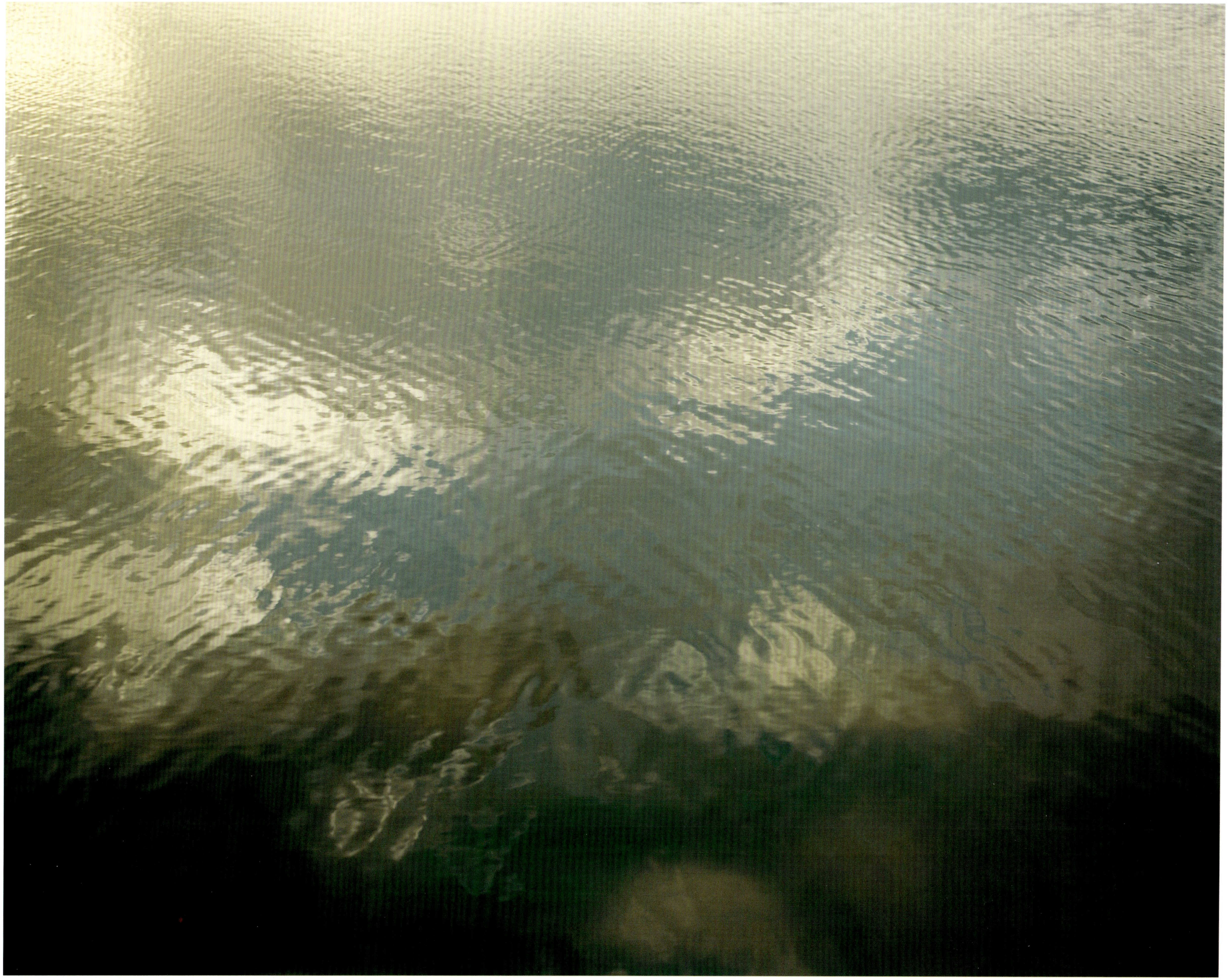

We wish to acknowledge the following individuals for their enthusiasm, participation, guidance and frequent assistance in bringing this book to fruition: Howard Stein and Joy of Giving Something, Chris Pichler, Mark Haworth-Booth, Maya Ishiwata, Alison Crosby, Jason Ma, Amy Lau, Mirta Rodriguez Gómez, Rob Friedman and Dave Read. A great many thanks to all of you! – Ed & Mirta

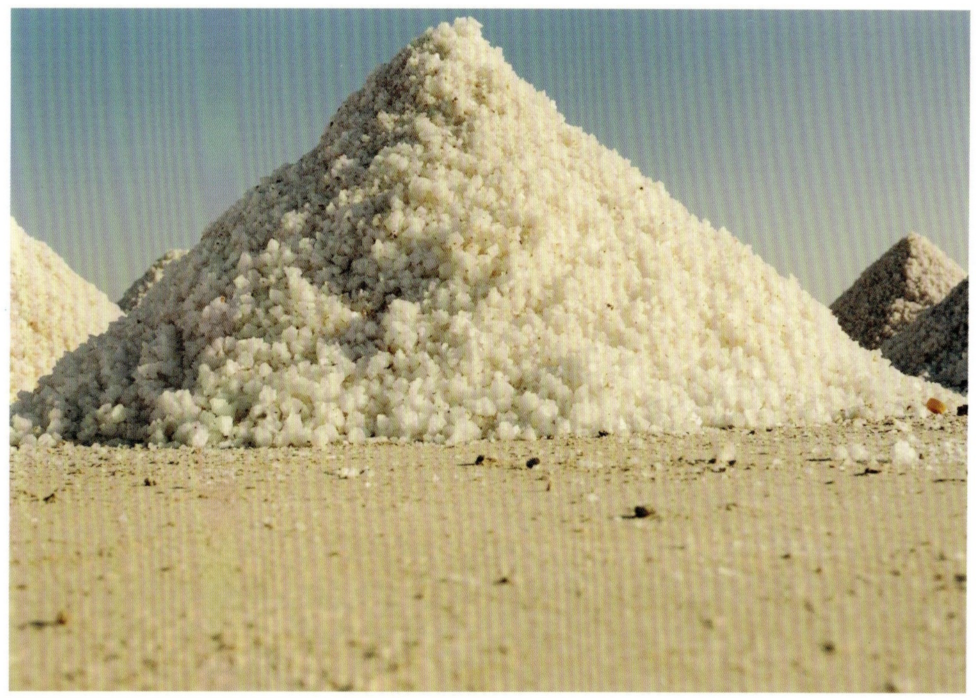

This first printing of *Fried Waters* is limited to 1,000 casebound copies, each with an original photograph, printed by the artists, on the front cover. A special edition of 25 copies with an additional 11x14-inch photograph, printed, numbered and signed by the artists and presented in a clamshell box, is available from the publisher. Printed in China for Nazraeli Press. ISBN 1-59005-090-8